802.11 DEMYSTIFIED

MCGRAW-HILL *DEMYSTIFIED* SERIES

802.11 Demystified

James LaRocca
Ruth LaRocca

McGraw-Hill

New York   Chicago   San Francisco   Lisbon
London   Madrid   Mexico City   Milan   New Delhi
San Juan   Seoul   Singapore   Sydney   Toronto

Cataloging-in-Publication Data is on file with the Library of Congress.

McGraw-Hill

A Division of The McGraw·Hill Companies

1 2 3 4 5 6 7 8 9 0 DOC/DOC 0 9 8 7 6 5 4 3 2

ISBN 0-07-138528-2

The sponsoring editor for this book was Judy Bass and the production supervisor
was Sherri Souffrance. It was set in Century Schoolbook by MacAllister Publishing
Services, LLC.

Printed and bound by R.R. Donnelley & Sons Company.

LIST OF TRADEMARKS AND COPYRIGHT HOLDERS

Institute of Electrical and Electronics Engineers (IEEE)—802.11 and 802.11b

Wireless Ethernet Compatibility Alliance (WECA)—Wi-Fi

Symbol—MobiLAN

HomeRF—OpenAir and SWAP

Intel—Springport and Anypoint

Proxim—RangeLAN, Symphony, and Harmony

Cisco—Aironet

Agere—WaveLAN and ORiNOCO

Microsoft—Windows and Windows XP

Apple Compute—AirPort

Compaq—MultiPort and iPAQ

Handspring—Visor

Texas Instruments—packet binary convolutional coding (PBCC)

To Olivia and Isabella, the joy in our lives, and to our families, who support all that we strive to become.

CONTENTS

INTRODUCTION

Wireless networking technologies such as 802.11, *third-generation* (3G) cellular, and Bluetooth are natural steps in the evolution of computers and telecommunications. Computers have become smaller, faster, and cheaper at an astounding rate. Initial computer installations were cumbersome pockets of isolated processing. If intercommunication was required, it was through manual methods such as punch cards or paper tape. In today's world of instantaneous high-speed communications, it's hard to believe that, not so long ago, one of the first telecommunication switching centers was the torn tape center. Here, paper tape computer transmissions were received via Teletype, read by an operator, and manually fed into the Teletype of the destination computer. The inevitable need to share data quickly and reliably led to the development of direct computer-to-computer communication technologies. Computer communications were initially implemented with a variety of proprietary schemes, and data networks remained private. Although manual processing had been lessened, the private, proprietary nature of computer communication limited its usefulness until the 1960s.

Early in that decade, Paul Baran of RAND began publishing a series of papers on communications networks, culminating in "On Distributed Communications" in 1964. This work introduced many of the concepts underpinning today's communications networks. Also introduced early in the 1960s was the first widely deployed digital transmission system. Called the *T-Carrier system*, it was created by the Bell System and gave us the terms T1, T3, and so on. In the late 1960s, the Department of Defense's *Advanced Research Projects Agency* (ARPA) implemented ARPANET. Utilizing packet-switching technology, ARPANET interconnected defense contractors, academic institutions, and government agencies. This network was the start of what we know today as the Internet.

The 1970s brought the creation and adoption of several key technologies. As the large monolithic mainframe gave way to local clusters of computers, we saw the segmentation of computer communications into the *local area network* (LAN) and the *wide area network* (WAN). Telecommunication standards began to evolve that addressed the unique needs of these two functions. At Harvard in 1973, Bob Metcalfe published his Ph.D. thesis on Ethernet. In 1974, *Bolt, Baranek, and Newman* (BBN) created the first public packet network. In 1974, Vint Cerf and Bob Kahn published the *Transmission Control Protocol* (TCP) as part of their work with ARPA. They continued enhancing the protocol, and in 1978 divided it into the *Internet*

Protocol (IP) for routing and the *Transmission Control Protocol* (TCP) for packet processing. The pieces were now in place for the computing and communications revolution of the 1980s.

With the advent of the personal computer in the early 1980s, desktop and home computing became a reality. The result was again isolated pockets of computing, but computers were now much smaller, in much larger quantities, and growing at an incredible rate. The ability to colocate a large number of small, powerful computers gave rise to a number of different LAN protocols. In addition to ARPA's TCP/IP set of protocols, several competing protocols were popularized. *Xerox Network Systems* (XNS) served as the basis for Banyan Vines and Novell NetWare. Apple had AppleTalk, and the *Digital Equipment Corporation* (DEC) had DECnet. These protocols were powerful individually, but presented an interoperability nightmare. WAN implementations were dominated by data networks employing such technologies as X.25 and IBM's System Network Architecture (SNA). Expensive to implement and hard to operate, they were ill suited to the communications needs of the new population of small, fast computers. There was clearly a need for a cheap, easy-to-use alternative for both LAN and WAN applications.

During this period of time, ARPANET continued to grow. Between 1980 and 1983, ARPANET converted its whole network to the TCP/IP set of protocols. In 1983, anyone wanting to connect to ARPANET was required to use TCP/IP. The *Defense Advanced Research Projects Agency* (DARPA) contracted BBN to implement a set of TCP/IP protocol stacks. The University of California at Berkley would distribute these free of charge with their implementation of the UNIX operating system. TCP/IP soon dominated UNIX networking. ARPANET eventually split off the military portion of its network into a separate network called MILNET. It combined the remaining research network with NSFNET, a WAN run by the *National Science Foundation* (NSF). This combined entity formed the backbone network for today's Internet.

The late 1980s saw the commercialization of this backbone network with the formation of companies like UUNET and PSINet who provided Internet access to businesses and individuals. TCP/IP became the protocol of choice for this network. The success of this network led to the adoption of Ethernet and TCP/IP as the networking protocol of choice for enterprises and LANs. Capitalizing on this trend, Cisco Systems was formed in 1986. Cisco's successful marketing of Ethernet equipment resulted in TCP/IP dominating the LAN marketplace. With computers still getting cheaper, smaller, and more powerful, the stage was set for Ethernet equipment and services to become a consumer product as well.

In the 1990s, companies such as Microsoft, Yahoo, America Online, and Netscape drove the growth of the Internet through advances in network access, information retrieval, and data presentation. The ability to easily share data between computers in a standard manner caused the Internet to grow from a government research network to a primary means of world-wide communication. Thousands of perfect strangers now form peer-to-peer networks like Napster or perform massive cooperative computing tasks like SETI@home. The networks of the new millennium connect millions of computers and operate at gigabit and soon terabit speeds.

One thing remained constant during this evolution from mainframe to personal computer—most of these devices couldn't move. A power cord and a wired network connection tethered them. If you wanted to move one, you powered it down, unplugged it, hauled it to its new home, and plugged it back in. This was clearly not aligned with the mobile, on-the-go world we had become. Advances in battery and power technology began to free the computer from its tethers. Laptops and notebooks emerged, but were just the initial realization of our need for mobile computing. We now have *personal digital assistants* (PDAs) in our briefcases and computers in our watches, vehicles, toys, appliances, and sometimes in our bodies. Although portable, these devices needed a physical connection for data networking until recently. Early attempts at providing a wireless connection to a data network were limited in speed, functionality, and reliability. As computers have become part of our everyday lives, the need for cheap, powerful, and flexible ways of exchanging data without being tethered has become a primary requirement for the development of the next wave of computing applications.

Up to this point, we have discussed computing and data networks, but this evolution is intertwined with the evolution of the telephone system. Historically, voice communication and data communication developed on separate paths. Although computer communication was digital from the start, telephone communication evolved as a product of the analog nature of the human voice. Voice networks and data networks employed architectures and protocols optimized for their specific payload. Advances in digital signal processing have changed the rules. The ability to digitize analog information efficiently has allowed the two worlds to merge. This is what has come to be known as a *converged* network.

Since Alexander Graham Bell's first call, the telephone network has grown into millions of miles of copper wire and optical fiber connecting the world's telephones. Arranged in a hierarchical network, switches route calls between endpoints. The *private branch exchange* (PBX) became the telephone equivalent of a local computer network. Telephone networks eventually adopted digital transmission (Bell Labs' T-Carrier) as the preferred transmission mechanism for the telephone network. In the last few years, optical fiber has become the physical medium at the heart of the network. Although the telephone network became digitized in many aspects, it still remained separate from the data network. It employed many protocols and transmission mechanisms that were still optimized for voice.

While the digitization of the voice network was taking place, computers began to require a network to exchange data. Because the existing voice network was accessible virtually everywhere and had become digitized in some respects, it only made sense that it could be utilized where possible for computer communications as well. Larger computers could provide a direct digital interface to the T-Carrier network. Smaller ones, like the PC, used a modem to make the computer appear to be a telephone. This worked fine as long as data traffic was a relatively small percentage of the traffic on the telephone network. But as we know, data transmission needs increased exponentially when the PC revolution spawned millions of PCs wanting a connection to the Internet. Engineered for voice traffic, the access portion of the traditional telephone network was overwhelmed. A network architecture where data traffic was offloaded onto an IP data network as soon as possible. Telephone companies began marketing voice services and data services.

As the Internet access demand grew, a new kind of specialized service provider was born—the *Internet service provider* (ISP). Created to provide Internet data access, ISPs can be divisions of the telephone company or independent entities like America Online. Technologies like *Digital Subscriber Line* (DSL) use the existing copper telephone connections to provide high-speed data as well as voice service. An alternative is provided by the cable television companies, who also provide Internet access through their networks. In fact, AT&T owns the largest cable system in the United States. Digital data communications is now a consumer product and primarily uses IP as the protocol of choice.

While digital data communications became a larger part of the existing telecommunications infrastructure, voice telephony was also undergoing major changes. Although the unique characteristics of data and voice gave rise to separate networks, recent advances in digital voice transmission have begun to bring the two technologies together. Voice is becoming just another type of data. The compelling economics of packet-based data networking has encouraged the migration of increasing amounts of voice traffic to packet data networks. This is referred to as *voice over IP* (VoIP). Still undergoing the growing pains of any new technology, it is predicted that eventually most voice traffic will be transmitted over packet networks.

Although end-to-end VoIP telephony is still a small percentage of voice traffic, it is increasingly likely that a call from your home phone may actually traverse an IP network before it reaches its destination. We have even seen the formation of companies providing voice services over the public Internet. The unpredictable nature of public Internet transmission makes this difficult until more sophisticated *quality of service* (QoS) mechanisms are widely available as part of the underlying transmission protocols.

The other major change in voice telephony has been the rapid growth of the cellular telephone industry. Just as computers became mobile in the 1980s, the need for mobility resulted in the creation of cellular phone networks. Today, cellular telephony is a primary means of voice communication across the globe. Voice communications became mobile, but data networks remained essentially wired. It was inevitable that the ubiquitous nature of computing, the convergence of voice and data networks, and the desire for mobility in our daily lives would lead to the commercialization of a variety of wireless data communication technologies. Cellular network carriers are implementing data-capable technologies such as *General Packet Radio Service* (GPRS), *Code Division Multiple Access 2000* (cdma-2000), *Universal Mobile Telecommunications Service* (UMTS), and *Enhanced Data for GSM Evolution* (EDGE). These technologies address wide-area data applications at speeds generally less than 2 Mbps. Emerging personal area network (PAN) technologies such as Bluetooth are targeted at short-range data applications.

The TCP/IP community has addressed the need for high-speed local area wireless access through the 802.11 series of specifications. Some of the same characteristics that allowed TCP/IP to become the dominant WAN/LAN technology are fueling 802.11's rapid adoption as a wireless network standard. This book explains where 802.11 fits into the wireless data picture, its technical underpinnings, and some of the trends that have sparked its explosive growth.

ACRONYMS

2G—Second-generation cellular

3G—Third-generation cellular

ADSL—Asymmetric Digital Subscriber Line

AES—Advanced Encryption Standard

AMPS—Advanced Mobile Phone System

ANSI—American National Standards Institute

AP—access point

ARPA—Advanced Research Projects Agency

ARPANET—Advanced Research Project Agency Network

bps—Bits per second

BFSK—binary frequency-shift keying

BPSK—binary phase-shift keying

BSS—basic service set

BSSID—Basic service set identification

CAT 5—Category 5 (cable)

CCA—Clear channel assessment

CCK—Complementary code keying

CDMA—Code Division Multiple Access

CDPD—Cellular Digital Packet Data

CMOS—Complementary metal oxide semiconductor

CRC—cyclic redundancy code

CS—carrier sense

CSMA/CA—carrier sense multiple access with collision avoidance

CSMA/CD—carrier sense multiple access with collision detection

CTS—clear to send

D-AMPS—Digital Advanced Mobile Phone System

DARPA—Defense Advanced Research Projects Agency

dB—Decibel

DBPSK—differential binary phase-shift keying

DCF—distributed coordination function

DECT—Digital European Cordless Telephone

DES—Data Encryption Standard

DFS—Dynamic frequency selection

DHCP—Dynamic Host Control Protocol

DIFS—Distributed (coordination function) interframe space

DNS—Domain Name Service

DOCSIS—Data over Cable Service Interface Specification

DPSK—Differential phase-shift keying

DQPSK—Differential quadrature phase-shift keying

DSL—Digital Subscriber Line

DSP—Digital signal processor

DSSS—Direct sequence spread spectrum

EDGE— Enhanced Data for GSM Evolution

EIA/TIA—Electronic Industries Association/Telecommunications Industry Association

EIFS—Extended interframe space

EIRP—Equivalent Isotropically Radiated Power

ESS—Extended service set

ETSI—European Telecommunications Standards Institute

FCC—Federal Communications Commission

FCS—Frame check sequence

FDMA—Frequency division multiplexing

FHSS—Frequency-hopping spread spectrum

FPGA—Field programmable gate array

FSK—Frequency-shift keying

FTP—File Transfer Protocol

GFSK—Gaussian frequency-shift keying

GHz—Gigahertz (billion cycles per second)

GPRS—General Packet Radio Service

GSM—Global System for Mobile Communications

HDSL—High-bit-rate Digital Subscriber Line

HIPERLAN—High-performance radio local area network

HomePNA—Home Phoneline Networking Alliance

HTTP—Hypertext Transfer Protocol

Hz—Hertz (cycles per second)

IAPP—Interaccess Point Protocol

IBSS—Independent basic service set

IEEE—Institute of Electrical and Electronics Engineers

IETF—Internet Engineering Task Force

IFS—Interframe space

IP—Internet Protocol

IR—Infrared

ISDN—Integrated Services Digital Network

ISM—Industrial, Scientific, and Medical (band)

ISO—International Standards Organization

ISP—Internet service provider

ITU—International Telecommunications Union

Kbps—Kilobits per second

KHz—Kilohertz (thousand cycles per second)

LAN—Local area network

LLC—Logical Link Control

LMDS—Local Multipoint Distribution Service

MAC—Medium Access Control (layer)

MAN—Metropolitan area network

Mbps—Million bits per second

MHz—Megahertz (million cycles per second)

MIB—Management Information Base

MLME—MAC layer management entity

MMAC—Mobile Multimedia Access Communication (Japan Promotion Council for)

MMDS—Multichannel Multipoint Distribution Service

MMPDU—MAC management protocol data unit

MPDU—MAC protocol data unit

MSDU—MAC service data unit

NAV—Network allocation vector

NDIS—Network Design Interface Specification

NIC—Network interface card

NIST—National Institute of Standards and Technologies

OSI—Open Systems Interconnection (model)

PAN—Personal area network

PBCC—Packet binary convolutional coding

PBX—Private branch exchange

PCF—Point coordination function

PCMCIA—Personal Computer Memory Card International Association

PDU—Protocol data unit

PHY—Physical (layer)

PIFS—Point (coordination function) interframe space

PLCP—Physical Layer Convergence Protocol

PLME—Physical layer management entity

PMD—Physical medium dependent

PPDU—PLCP protocol data unit

PRNG—Pseudorandom number generator

QAM—Quadrature amplitude modulation

QFSK—Quadrature frequency-shift keying

QoS—Quality of service

RF—Radio frequency

RSA—Rivest, Shamir, and Adleman public-key cryptosystem

RTS—Request to Send

SDR—Software-Defined Radio

SDSL—Synchronous Digital Subscriber Line

SDU—Service data unit

SIFS—Short interframe space

SiGe—Silicon germanium

SNA—Systems Network Architecture

SoC—System-on-a-chip

SWAP—Shared Wireless Access Protocol

TCP—Transmission Control Protocol

TDMA—Time Division Multiple Access

TPC—Transmission power control

UMTS—Universal Mobile Telecommunications System

URL—Uniform Resource Locator

USB—Universal serial bus

VDSL—Very high rate Digital Subscriber Line

VoIP—Voice over IP

VPN—Virtual private network

W-CDMA—Wideband CDMA

WAN—Wide area network

WAP—Wireless Application Protocol

WECA—Wireless Ethernet Compatibility Alliance

WEP—Wired Equivalent Privacy

Wi-Fi—Wireless Fidelity

WISP—Wireless Internet service provider

WLAN—Wireless local area network

WLL—Wireless local loop

WPAN—Wireless personal area network

WPBX—Wireless private branch exchange

XNS—Xerox Network Systems

802.11 DEMYSTIFIED

802.11— Bridging the Gap

Wireless communication is changing the way we work and play. From light-based methods such as lasers and infrared to radio-based *local area networks* (LANs), *personal area networks* (PANs), and cellular telephones, we are being liberated from the limitations of a wired world. All these technologies have existed in some form for many years, but until recently, widespread commercial adoption has been slow. The most notable exception has been cellular telephony because of its focus on voice transmission. By 1980, telephones were ubiquitous, but few people had a computer. As the demand for mobility grew, it was natural that voice communication would be the first to be unwired, but mobile data communication capabilities remained limited at best. The *personal computer* (PC) revolution and the explosive growth of the Internet have made general-purpose computers as common as the telephone. We now have households with multiple computers and workers frequently use laptops or *personal digital assistants* (PDAs) in addition to their desktop PCs. With embedded computers now designed into everything from car engines to children's toys, the possibilities for interconnection are staggering. In some cases, these systems communicate over traditional wired media, but their full potential is only now being unlocked as they are freed from physical interconnection.

With such a wide range of possible applications, it is understandable that a number of different wireless data communication solutions would arise. These include the *Institute of Electrical and Electronics Engineers* (IEEE) 802.11 standard, *high-performance radio local area networks* (HIPERLANs), *home radio frequency* (HomeRF), Bluetooth, *General Packet Radio Service* (GPRS), *third-generation* (3G) cellular, and *free-space optics* (FSO). With overlapping functionality, a seemingly endless stream of technical variants, and differing states of commercial maturity, it seems a daunting task to fit them together. For the reasons examined in the following sections, the 802.11 standard has emerged as a key enabler for the growing universe of wireless data applications. This chapter provides an overview of key wireless technologies and describes how 802.11 solutions bridge the gap between other wireless communication standards.

A convenient way to begin organizing wireless data networking solutions is by distance (see Figure 1-1). Although longer-distance technologies would seem to provide the same capabilities as shorter-

Figure 1-1
The wireless universe

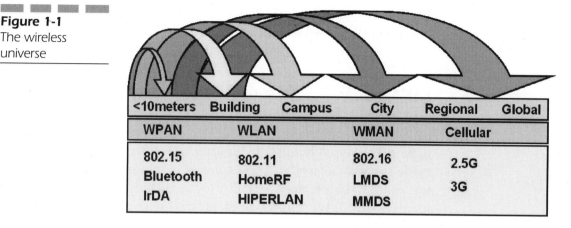

Figure 1-1
The wireless universe

<10meters	Building	Campus	City	Regional	Global
WPAN	WLAN		WMAN	Cellular	
802.15	802.11		802.16	2.5G	
Bluetooth	HomeRF		LMDS	3G	
IrDA	HIPERLAN		MMDS		

distance technologies, factors such as frequency availability, transmission speed, interference, cost, and government regulation have resulted in specific applications for each technology. A short-range wireless network with communication distances of 10 meters or less has come to be known as a *wireless personal area network* (WPAN). The most notable WPAN is Bluetooth. The IEEE 802.15 Working Group is developing standards for WPANs. A *wireless local area network* (WLAN) generally handles distances of several hundred meters or less and typically links devices within a building or campus. In this group, we have IEEE 802.11, HIPERLAN, and HomeRF. A *wireless metropolitan area network* (WMAN) provides broadband wireless access for small regional areas such as cities or metropolitan areas. Wireless MAN protocols include *Local Multipoint Distribution Service* (LMDS) and *Multichannel Multipoint Distribution Service* (MMDS).

The IEEE 802.16 Working Group develops standards and recommended practices for WMANs. Large regional and global wireless data networks are in the domain of cellular telephone networks. GPRS, IS-95B, and *Enhanced Data for GSM Evolution* (EDGE) are examples of cellular networks that provide data capabilities as an extension of *second-generation* (2G) cellular networks. These are referred to as 2.5G cellular standards. The *International Telecommunication Union* (ITU) has published a framework for 3G networks called the *International Mobile Telecommunications 2000 Project* (IMT-2000). It includes wireless standards such as *Universal Mobile*

Telecommunications Service (UMTS), *Wireless Code Division Multiple Access* (WCDMA), and CDMA-2000.

802.11 WLANs—Why Now?

LANs were originally created as short-distance data communications networks. Advances in routing and transmission speed have made the local aspect of LANs somewhat of a misnomer. Devices on the same network may actually be thousands of miles apart, but seem nearby. LANs can be implemented with a variety of network topologies (such as Token Ring and Ethernet), data communication protocols (such as the *Transmission Control Protocol/Internet Protocol* [TCP/IP] and IBM's *Advanced Program-to-Program Communication* [APPC]), and transmission media (such as twisted-pair wire, coaxial cable, optical fiber, and radio). A WLAN simply uses radio or light as the preferred transmission medium and is generally restricted to an area more like the first wired LANs—that is, they generally link a set of devices in a room, building, or group of closely located buildings. Until recently, WLANs were viewed as a costly solution with limited scope. Several factors have recently brought WLANs to the forefront as a viable and, in some cases, preferred option for local area networking. Telecommunications researcher Joseph Pelton states that the readiness for adoption of a telecommunications technology is indicated by advancement in five key areas: standards, regulation, technology, services, and economics. In the last few years, WLANs have seen significant advancements in all five areas, leading to today's increasing rate of market acceptance. This is particularly true of the IEEE 802.11 series of WLAN specifications. Let's examine the advances in all five areas.

802.11 Standards—The Family Tree

Standardization and interoperability are key factors in the rapid growth of 802.11. Three complementary standardization initiatives underlie the recent acceptance of 802.11. The first is the standardization of LANs by the IEEE Project 802. The second is the worldwide acceptance of the IP suite through the efforts of the *Internet Engineering Task Force* (IETF). The third is the very successful interoperability initiative of the WLAN industry itself—the *Wireless Ethernet Compatibility Alliance* (WECA).

The IEEE began meeting in 1980 to discuss LAN and MAN standards. This project was given the designation 802 and in 1990 the IEEE formally approved the 802 standard. The 1990 IEEE 802 standard provides the framework for organizing IEEE LAN/MAN standards. IEEE 802 is based on the *Open Systems Interconnection* (OSI) model of the *International Standards Organization* (ISO). The OSI model segments a data communication protocol into seven distinct functional layers: physical, data link, network, transport, session, presentation, and application. The physical layer specifies the physical properties of a medium including electrical, optical, or radio characteristics. The data link layer describes how the transmitted data is organized into logical groups such as packets, frames, or blocks. It also includes error checking for possible physical media errors. The network layer describes how multiple transmissions are grouped into logical exchanges between entities and includes the network address scheme. The transport layer specifies the characteristics of data delivery including the sequencing of data blocks and error mechanisms like retransmission. The session layer describes how lower-level data blocks or packets are grouped into sequences or dialogs for specific functions. An example is the definition of request and reply sequences. The presentation layer describes the syntax of the data being transferred such as how alpha or numeric characters are represented. The application layer describes the protocol for a specific application, such as database access or file transfer. Although the OSI model has proved very useful for defining the basic functions of communications protocols, individual protocols, such as IEEE 802 or the IP suite, may not map directly to each layer. Some layers may be joined together or further subdivided.

The IEEE 802 framework defines the lowest two layers of the OSI reference model—the physical and data link layers. The IEEE 802 physical interface layer corresponds directly to the OSI physical layer and is referred to as the PHY layer. The IEEE 802 PHY layer defines the physical characteristics of the transmission media (such as twisted-pair or coaxial cable) and the physical signaling protocol required for putting data on the media. In order to handle the unique requirements presented by a shared access LAN, the IEEE 802 Committee segmented the OSI data link layer into two components—the *Medium Access Control* (MAC) layer and the *Logical Link Control* (LLC) layer. The MAC layer defines the protocol for accessing the PHY layer, whereas the LLC layer enables multiple-user access to the MAC.

When looking at the IEEE 802 standards, multiple PHYs and a MAC may be grouped together for a specific network access method. The best-known example of this is IEEE 802.3, which specifies a network access method using a technique called *carrier sense multiple access with collision detection* (CSMA/CD). It should be noted that Ethernet is another very similar CSMA/CD network access method. Ethernet and IEEE 802.3 tend to be used interchangeably, but technically 802.3 is an IEEE standard and Ethernet is the protocol invented by Bob Metcalfe and commercialized by Xerox in 1973. Two of the more familiar PHY standards contained in IEEE 802.3 are 10Base-T and 100Base-T. These support 10 and 100 Mb transmission over twisted-pair cable. Other common network access protocols in the IEEE 802 family are IEEE 802.4 (token bus) and IEEE 802.5 (Token Ring). Ethernet, Token Ring, and token bus are all wired LAN access methods. Wireless LANs are specified in the IEEE 802.11 series of standards and have their own unique PHY and MAC layers.

The LLC layer is specified in IEEE 802.2 and is independent of network access specifications. It provides a common method for establishing and maintaining a logical communication link without specific knowledge of the network access method or physical medium. It is the gateway to the higher-layer protocols discussed in the following section.

While the IEEE 802 Committee was standardizing the PHY and data link layers of the OSI model for local area networks, the *Advanced Research Project Agency Network* (ARPANET) and the IETF were actively engaged in standardizing and promoting the IP suite. The IP suite contains protocols and recommendations that loosely correspond to the upper five layers of the OSI model (see Figure 1-2). These include the User Datagram Protocol (UDP), the Transmission Control Protocol (TCP), and the Internet Protocol (IP). The standardization of low-level network access by the IEEE and the

Figure 1-2
OSI, 802, Internet
protocol mapping

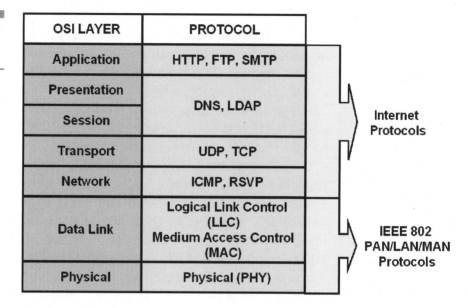

OSI LAYER	PROTOCOL
Application	HTTP, FTP, SMTP
Presentation	DNS, LDAP
Session	DNS, LDAP
Transport	UDP, TCP
Network	ICMP, RSVP
Data Link	Logical Link Control (LLC) Medium Access Control (MAC)
Physical	Physical (PHY)

Internet Protocols

IEEE 802 PAN/LAN/MAN Protocols

widespread adoption of the IP suite as a preferred higher-level communication method set the stage for the emergence of 802.11 as the dominant WLAN technology.

With IEEE 802.3 WLANs widely deployed and wireless technology finding its way into everything from cordless phones to retail checkout counters, WLAN standards were a logical addition to the IEEE 802 family. In 1997, the IEEE published its first standard for WLANs—IEEE 802.11. IEEE 802.11 supports three wireless PHY interfaces operating at speeds of 1 and 2 Mbps. The first IEEE 802.11 PHY uses infrared as the transmission medium. This has never been widely adopted. The other two use unlicensed 2.4 GHz radio in the *Federal Communications Commission's* (FCC's) *Industrial, Scientific, and Medical* (ISM) band, but employ different methods of spread spectrum transmission. The second IEEE 802.11 PHY employs *frequency-hopping spread spectrum* (FHSS) in the 2.4 GHz ISM band. The third 802.11 PHY employs *direct-sequence spread spectrum* (DSSS) in the 2.4 GHz ISM band.

Even with the publication of 802.11 in 1997, the market was still slow to commercialize WLAN systems. Operating at speeds of 1 and 2 Mbps, WLANs were slow compared to their 802.3 counterparts, radio technology was costly to produce, and optional features like encryption and power management caused interoperability issues. In addition, semiconductor technology for 2.4 GHz radios was still maturing, keeping basic material costs high enough to prevent widespread adoption. This began to change in 1999 as the market potential of WLANs began to be realized. Led by companies like Lucent (now Agere) and Harris (now Intersil), 2.4 GHz semiconductor radio technology had matured enough for basic materials costs to drop. The speed issue was addressed in 1999 with the publication of two Radio PHY amendments to IEEE 802.11. The first, IEEE 802.11b, added support for data rates of 5.5 and 11 Mbps in the 2.4 GHz band. In 1999, it was the fastest 2.4 GHz IEEE 802 WLAN standard and was sometimes referred to as IEEE 802.11 *high rate* (HR). This name is rarely used now and the approval of the 2.4 GHz IEEE 802.11g (22 Mbps) in 2001 made IEEE 802.11b the second fastest 2.4 GHz WLAN.

The second PHY amendment in 1999 was IEEE 802.11a. IEEE 802.11a is fundamentally different from IEEE 802.11b in that it uses 5 GHz radio spectrum, utilizes a radio transmission technique called *orthogonal frequency division multiplexing* (OFDM), and supports data rates up to 54 Mbps. Although its data rates are significantly higher, 5 GHz semiconductor radio technology is harder to produce and this technology is just now maturing. It is where 2.4 GHz semiconductor radio technology was in 1999, and if it follows a similar path, IEEE 802.11a systems will see substantial growth in 2002 and 2003.

WLANs became more than a niche product with the standardization of higher speeds and the maturation of 2.4 GHz radio technology in the late 1990s. Component and equipment costs had begun to drop and 11 Mbps was competitive with WLANs, but interoperability concerns still held back widespread IEEE 802.11b deployment. To address this problem, major vendors formed the Wireless Ethernet Compatibility Alliance (WECA). Formed by 3Com, Aironet (acquired by Cisco), Intersil, Lucent Technologies, Nokia, and Symbol Technologies in 1999, WECA has grown to nearly 150 members in 2002. This growth in WECA membership not only shows the explosive growth of IEEE 802.11b, but it is also a testament to the power of standards and interoperability as a key factor in the adoption of any technology. WECA produced an interoperability benchmark for 802.11b and began issuing certifications to equipment vendors in April 2000. As part of this effort, WECA established the Wi-Fi™

brand. Wi-Fi stands for Wireless Fidelity. Standardized interoperability testing ensured that customers buying IEEE 802.11b products with the Wi-Fi seal of approval could be confident that they would work together. In 2001, WECA announced they would begin interoperability testing of 5 GHz IEEE 802.11a products and established the Wi-Fi5™ seal of approval for these products.

802.11—No License . . . No Problem

Pelton's second criterion for communications technology adoption is regulation. A regulatory environment conducive to the commercialization of IEEE 802.11 technology has been the result of unlicensed spectrum access and the active efforts of standards bodies and regulatory agencies to provide an organized framework without stifling innovation. Although the FCC regulates all North American radio spectrum in some respect, it has divided it into licensed and unlicensed categories. In the licensed bands, the government restricts the number of participants. Direct government oversight of the terms of the license and the license fees themselves make it much more costly to use. Cellular telephone carriers, radio stations, and television broadcasting all use licensed spectrum. Licensing ensures a much more orderly use of the spectrum, but the high costs, financial defaults, lawsuits, and implementation delays associated with recent cellular spectrum auctions are examples of the downside.

On the other hand, the unlicensed bands are free to any participant as long as they adhere to the FCC's basic rules for use. This promotes widespread use, but can lead to problems with the overpopulation of the frequency and interference. Because the IEEE 802.11 series of WLANs operate in the unlicensed 2.4 and 5 GHz radio bands, we have seen both sides of that coin. The same frontier attitude that helped drive the growth of the LAN industry and the Internet has driven the rapid adoption of IEEE 802.11 WLANs. At the same time, concerns over interference and decreased performance have been raised. The most publicized conflict is between Bluetooth and 802.11b in the 2.4 GHz band. Although a concern, it has not appreciably slowed the adoption of either technology yet. Equipment vendors, the IEEE, and the FCC are actively addressing this issue, which we'll cover later in more detail.

802.11—Smaller, Cheaper, and Faster

Pelton's third criterion is technological readiness. WLANs have been the beneficiary of general advances in semiconductor materials technology that has produced lower-cost, higher-performance chips. Many of the advances in the new compound semiconductor materials are being applied in the wireless arena. A good example of this is the development of *silicon germanium* (SiGe) chips. Pioneered by IBM, SiGe chips are relatively easy to process and have excellent noise characteristics. With applications in intermediate *radio frequency* (RF) chips and low-noise amplifiers, the efficiencies gained through advances in these basic building blocks have translated into simpler, less costly components.

A great deal of research has also been done to optimize use of the limited bandwidth allocated by regulatory agencies for WLANs. Sophisticated spread spectrum modulation techniques for both FHSS and DSSS have found applications in IEEE 802.11 series products. These techniques have helped overcome interference while increasing speed and range. IEEE 802.11 originally supported two modulation schemes for DSSS and two modulation schemes for FHSS. The direct sequence schemes were based on *differential phase-shift keying* (DPSK). The frequency-hopping schemes were based on *Gaussian frequency-shift keying* (GFSK). These proved inadequate for the 5 and 11 Mbps rates of 802.11b. To address this need, IEEE 802.11b adopted two new modulation techniques— *complementary code keying* (CCK) and *packet binary convolutional coding* (PBCC). For IEEE 802.11a (5 GHz, 54 Mbps), OFDM was chosen. The 22 Mbps speeds of IEEE 802.11g are enabled by both OFDM and PBCC radio transmission techniques.

802.11—Services with a Smile

The fourth Pelton criterion is services. Because IEEE 802.11 was designed as a complementary replacement for the PHY and MAC layers of 802.3, services and applications designed for 802.3 can easily use 802.11 with little or no modification. This meant that as 802.11 matured, the vast array of deployed 802.3 services provided an immediate application market. Commercial applications using standard IP like the *File Transfer Protocol* (FTP) and *Hypertext Transfer Protocol* (HTTP) were immediately enhanced with wireless connectivity as an option. Contrast this with 3G cellular whose adoption has been slowed in large part by the lack of a "killer app" to drive end user demand.

The demand for access to the Internet regardless of location spawned public WLANs. Carriers like Wayport are currently deployed in airports, hotels, and convention centers across the globe. Starbucks is deploying 802.11b wireless access in over 4,000 stores by 2003. With an established applications market providing a base for growth, we are now seeing the development of applications and services specifically for the WLANs. The marriage of WLAN connectivity with relatively cheap, powerful devices such as the PDA are enabling applications in areas as diverse as manufacturing, education, medicine, retail, and the military.

802.11—Economics

All of the previous factors contributed to the satisfaction of Pelton's fifth criterion—a favorable economic model. With a ready market for services as well as reduced component costs, most major computer, telecommunication equipment, and semiconductor manufacturers have invested in the 802.11 market in some manner. IBM, Dell, Intel, Microsoft, Texas Instruments, Nokia, Agere, and Cisco (just to name a few) have made major investments in WLAN technology. The favorable economics of IEEE 802.11-based WLANs has resulted in equipment sales of over $1 billion in 2000 with a projected market of $3 billion in 2002. It is undeniable that 802.11 has arrived and is playing an increasingly important role in how we access the world of data surrounding us.

802.11—What's in a Name?

The explosive growth of the 802.3 and 802.11b industries following the publication of these standards illustrates how standards can serve as a catalyst for the adoption of a technology. To take advantage of this dynamic, it is important to have a basic understanding of the workings of the IEEE standards groups and procedures. The IEEE is involved with setting standards in a wide variety of fields and has a robust set of policies, procedures, and organizations to investigate technology and reach consensus among the membership on standards. This often results in cryptic names for what may be fairly commonplace things. The commercially spawned HomeRF is descriptive and Bluetooth is whimsical, but 802.11b falls somewhat short of stirring the imagination. The alphanumeric soup of the IEEE standards can be somewhat disorienting, but there is a method to the names. In order to provide a frame of reference, let's examine the IEEE standards process and look at each member of the 802.11 family.

The IEEE is a global professional organization with over 300,000 members and is involved with a number of activities in addition to setting standards. The IEEE is made up of 10 regions and 36 technical societies. The *IEEE Standards Association* (IEEE-SA) administers the standards process within the IEEE. The *American National Standards Institute* (ANSI) accredits the IEEE-SA and oversees the standards-setting process. This leads to many IEEE standards becoming ANSI standards as well. The standards process is one of open proposal, discussion, and consensus. When the IEEE-SA initiates a new technical investigation, it assigns the investigation to a sponsor in a related technical society. The Computer Society is responsible for the IEEE 802 series of standards through its LAN/MAN Standards Committee. IEEE 802 is simply the numerical designation given to the LAN/MAN Standards Committee when it was formed in 1980.

IEEE 802 LAN/MAN Standards Committee

Although the original charter of the IEEE 802 Committee was for local and regional area networks, its charter now also includes personal area networks. The committee's work is organized in a hierarchical fashion in which each level is responsible for a more specific aspect of the overall task. The main components of the IEEE 802 Committee are Working Groups and Technical Advisory Groups. The main difference between Working Groups and Technical Advisory

Groups is that Technical Advisory Groups do not write standards. They are responsible for giving guidance and making recommendations to the Working Groups on topics that require specialized expertise. Working Groups and Technical Advisory Groups are also given numerical designations. The IEEE 802 Wireless LAN Working Group was assigned 11—hence, the name IEEE 802.11. A Working Group with no current projects becomes inactive and may even be disbanded if its work has been superseded or completed. Table 1-1 shows the IEEE 802 Committee's Working Groups and Technical Advisory Groups.

Working Groups and Technical Advisory Groups may also contain Task Groups for even more specialized work. Task Groups are given

Table 1-1

LAN/MAN
Standards
Committee
Working Groups

802.0	Sponsor Executive Committee (SEC)
802.1	High-Level LAN Interfaces Working Group
802.2	Logical Link Control Working Group (inactive)
802.3	CSMA/CD (Ethernet) Working Group
802.4	Token Bus Working Group (inactive)
802.5	Token Ring Working Group (inactive)
802.6	MAN Working Group (inactive)
802.7	Broadband Technical Advisory Group (inactive)
802.8	Fiber Optics Technical Advisory Group (disbanded)
802.9	Isosynchronous LAN Working Group (inactive)
802.10	Interoperable LAN Security Working Group (inactive)
801.11	WLAN Working Group
802.12	Demand Priority Working Group (inactive)
802.14	Cable Broadband Comm. Network Working Group (disbanded)
802.15	Wireless Personal Area Network Working Group
802.16	Broadband Wireless Access Working Group

a letter designation. The Task Group formulating the standard for WLAN data rates of 5.5 and 11 Mbps was given the designation *b* resulting in what we know as IEEE 802.11b. An interesting twist of this system is that a name like IEEE 802.11a would seem to imply a precursor to IEEE 802.11b. In fact, they are related, but independent, specifications. One other entity called a Study Group may be formed to explore special topics. No special naming is used for Study Groups. Table 1-2 shows the Task and Study Groups within the IEEE 802.11 Working Group.

Table 1-2

802.11 Task and Study Groups

802.11	Wireless LAN PHY and MAC Specification (infrared and 2.4 GHz radio) Completed
802.11a	Wireless LAN PHY and MAC Specification for the 5 GHz Radio Band Completed
802.11b	Higher-Speed (5.5 and 11 Mbps) Wireless LAN PHY and MAC for 2.4 GHz Radio Completed
802.11c	Bridge Operation with IEEE 802.11 MACs (incorporated into 802.1D) Completed
802.11d	Extensions to 802.11 for Operation in Additional Regulatory Domains Completed
802.11e	802.11 MAC Quality of Service for Advanced Applications (for example, voice, video, and audio) Originally included security and authentication, now part of Task Group *i* Active
802.11f	Multivendor Access Point Interoperability Across Distribution Systems *Interaccess Point Protocol* (IAPP) Active
802.11g	Higher-Rate (20+ Mbps) Extensions in the 2.4 GHz Band Active

Table 1-2 *cont.*

802.11 Task and
Study Groups

802.11h	Enhancements for Dynamic Channel Selection and Transmit Power Control Active	
802.11i	Enhancements for Security and Authentication Active	
5 GHz Study Group (5GSG)	5 GHz Globalization. Joint study with the *European Telecommunications Standards Institute/Broadband Radio Area Network* (ETSI/BRAN) and *Mobile Multimedia Access Communication* (MMAC) (Japan) to promote interworking among the 802.11a, ETSI HIPERLAN/2, and MMAC standards. Active	

Understanding the IEEE Standards Process

Now that we have an outline of the IEEE 802 organization, let's examine exactly how standards become approved and published. The *Sponsor Executive Committee* (SEC) or Working Group 802.0 administers the IEEE 802 LAN/MAN Standards Committee. This group is responsible for the general administration of LAN/MAN standards activities and oversees the other IEEE 802 Working Groups. If someone is interested in investigating an area that may require standardization, they submit a *Project Authorization Request* (PAR) to the IEEE 802 SEC. A PAR not only describes the scope of the project, but it also must provide background material on how a standard in this area meets the IEEE 802 Committee's criteria for establishing a standard. These criteria include technical feasibility, economic feasibility, market potential, and IEEE 802 compatibility. The SEC votes on the PAR. If approved, it assigns the project to a new or existing Working Group and appoints a chairperson. Each chairperson is responsible for overseeing the activities of the group and setting the operating procedures based on the specific work of the group. Individuals may become members of a Working Group by meeting minimum attendance standards. Although encouraged, it is not necessary to be a member of the IEEE to be a member of an IEEE 802 Working Group. It is also important to note that members participate as individuals, not as formal representatives of business, professional, or technical organizations. Nonetheless, many major contributors to the IEEE standards process have a commercial affiliation and at times the standards proposal process may mirror commercial conflicts.

Voting and Approval Within the Working Group, Task Groups create draft standards for specific topics (for example, security). Proposals for the standards are submitted to the responsible group, which will discuss, revise, and vote on the proposed standard until consensus is reached on a final draft version. The final draft version is submitted for a membership vote. At this point, 75 percent of the membership must approve the draft before it can be sent to Sponsor Executive Committee for further action. The draft is revised, circulated, and voted upon until a 75 percent majority is achieved. Any member who votes No must indicate what changes should be made to change the vote to Yes. This mechanism is meant to lead the group into a cycle of constructive criticism that helps validate the standard when finally approved by the membership. Once consensus is reached, the draft standard is sent to the SEC for approval and transfer to the IEEE Standards Review Committee. The IEEE Standards Review Committee's approval allows it to be published as an IEEE standard. Around the same time, it may also be sent to the International Organization for Standardization for approval and publication as an international standard. Although the whole process from investigation to publication may sometimes be lengthy, the open forum and consensus approach results in broad-based standards with commercial viability. Those standards demanded by the marketplace and showing near-term commercial promise (like 802.11b) can be approved without an inordinate amount of delay. If an organization or business is considering implementing or producing IEEE 802.11 products, following and participating in the activities of the IEEE 802.11 Working Group is essential to understanding where the various flavors of 802.11 fit into the marketplace.

The IEEE 802.11 Alphabet Spelled Out

Let's take a look at the *Institute of Electrical and Electronics Engineers* (IEEE) 802.11 Working Group and its Task Groups in more detail. Around the beginning of 1996, the IEEE 802 LAN/MAN Standards Committee formed a Working Group to investigate the standardization of *local area networks* (LANs) implemented with wireless connectivity. This Working Group was given the designation 11, resulting in the name IEEE 802.11. Operating within the IEEE 802 Committee's well-established framework for producing LAN standards, the IEEE 802.11 Working Group set out to define wireless *physical* (PHY) and *Medium Access Control* (MAC) layers that would coexist as a peer to the IEEE 802.3 wired PHY and MAC layers. Several issues are introduced by the fundamental differences between wired and wireless connections.

The IEEE 802.11 Working Group— Going Mobile

The IEEE 802.11 Working Group was tasked with devising a standard that would not only support devices in a fixed location, but would also support portability and mobility. The difference between portability and mobility is important. Portability is a device's capability to be moved easily from place to place, but once a location is established, the device remains in a fixed position. Wired devices provide some degree of portability, but are restricted by the availability of power and a physical network connection. Before the advent of battery-powered computers, this meant that when a device was moved, it generally had to be powered off and the network connection was lost, unless of course, you were moving a very short distance or had very long LAN and power cables. Battery-powered devices removed the restriction for a physical power connection, but the network still had to be unplugged and network capabilities were lost during the move.

An example of portability in the wired world is moving a computer to the next office or carrying a laptop into a conference room. Wired portability assumes that the destination has enough power and physical network connections to support the devices being moved. This has obvious problems. It restricts the destinations to predetermined locations, and engineering these locations for the greatest number of users can be quite costly and inefficient. Have you ever gone into a conference room with two LAN outlets and five people with laptops or had to navigate through a thicket of LAN cables as you made your way to that last seat in the back? Portability in the wireless world means that 10 people can pick up their *personal digital assistants* (PDAs) or laptops and hold meetings anywhere in the area served by their *wireless LAN* (WLAN).

Mobility, on the other hand, implies that a device may be in motion while in use. In order to maintain full functionality, it must not only have a self-contained power source, but it must also maintain network connectivity while moving. In the wireless world, devices in constant motion can maintain network connections without stopping. Bicycle racers can transmit biometric data to a trainer as they speed around a track, or remote sensors can transmit real-time telemetry data to a control system as they move about.

In order to support fixed, portable, and mobile applications, the IEEE 802.11 Working Group had to first decide on the appropriate signal transmission media. They chose to support radio waves and infrared light. The basic differences between infrared light and radio waves are that radio waves broadcast their signals over an area, pass through most objects in their path, and do not require a clear line of sight.

There are two types of infrared systems: direct and reflective. *Direct infrared* is more suited to point-to-point applications over short distances. *Reflective* systems may cover a wider area, but are limited to operation within an enclosed area. Using infrared to achieve coverage equivalent to a radio-based system requires much more equipment and is generally not cost effective. Although infrared is supported in the IEEE 802.11 specification, it has not been widely implemented as a WLAN solution. Radio has become the medium of choice as evidenced by the three radio-based descendents of the original IEEE 802.11: IEEE 802.11a (5 GHz, 54 Mbps), IEEE 802.11b (2.4 GHz, 11 Mbps), and IEEE 802.11g (2.4 GHz, 22 Mbps).

IEEE 802.11—Challenges and Hurdles

Although radio is well suited for WLAN implementations, it comes with some inherent challenges that must be addressed: radio signal optimization, mobility management, security, and international regulatory coordination. The 1997 IEEE 802.11 standard and its 1999 revision addressed each of these areas in some manner. As deployments accelerated and the technology evolved, each of these areas required enhancement. This led to the formation of Task Groups investigating enhancements. As detailed in Table 1-2, some of these Task Groups have finished their work and others are still active. Task Groups IEEE 802.11a, IEEE 802.11b, and IEEE 802.11g address the need for greater transmission speeds and optimized *radio frequency* (RF) usage. Task Group IEEE 802.11e is tasked with examining application implementation issues including traffic prioritization and *quality of service* (QoS). As security became more of a concern, this investigation was removed from Task Group IEEE 802.11e and Task Group IEEE 802.11i was formed to concentrate on this area. Task Groups IEEE 802.11c and IEEE 802.11f work on WLAN interconnectivity and mobility protocols, respectively. Finally, Task Groups IEEE 802.11d, IEEE 802.11h, and the *5 GHz Study Group* (5GSG) look at international regulatory issues.

IEEE 802.11—Why 2.4 GHz?

Because radio signals are part of and interact with the environment within which they operate, they are subject to a variety of conditions that affect their quality, performance, and range. Physical characteristics such as building materials, the placement of objects within a coverage area, and other radio sources affect the propagation of a radio signal and can cause conditions such as interference, fading, and multipath. There is also a natural relationship between signal strength, transmission rate, and distance. Generally speaking, at any given signal strength, as the signal distance increases, the transmission rate decreases. In addition, lower frequencies are able to travel longer distances without degradation, but they carry less information.

Many of the design decisions made in the IEEE 802.11 PHY layer were directed toward the optimal use of the radio spectrum. Because IEEE standards are produced with commercial viability in mind, one of the first decisions facing the IEEE 802.11 Committee was deciding on the appropriate radio spectrum to use. It had to be a frequency that was available for general use, could be implemented relatively cheaply, and would support the data rates required for WLAN applications. Because radio spectrum is regulated worldwide, certain frequencies are allocated and unavailable, some are subject to licensing restrictions and costs, and others are unlicensed and available to anyone. All are subject to equipment regulations on output power and interference. The regulation of radio spectrum crosses geographic and political boundaries, and equipment regulations can have implications for any technology aspiring to worldwide deployment.

In North America, the *Federal Communications Commission* (FCC) has allocated unlicensed radio spectrum for general use. This spectrum is known as the *Industrial, Scientific, and Medical* (ISM) band. Some of the more commonly used bands are 900 MHz and 2.4 GHz. The 2.4 GHz band is generally regulated and used worldwide in a similar manner (for example, for cordless phones and microwave ovens). With this in mind, the IEEE 802.11 Working Group settled on 2.4 GHz as the frequency band that provided the best solution for low-cost, widely accessible, high-speed WLAN devices.

IEEE 802.11—2.4 GHz Spectrum Regulation

Devices in the 2.4 GHz band are subject to regulation by the FCC in the United States and by other agencies across the globe. The actual band specified by the FCC ranges from 2.4 to 2.4835 GHz. The FCC rules for radio devices are specified in Part 15 of Title 47 of the Code of Federal Regulations. These rules specify things such as emissions, output power, and spectrum management techniques. The Part 15 rules have a direct effect on the implementation of the IEEE 802.11 PHY layer. FCC ISM devices are required to operate at a maximum of 1,000 mW, although most operate at much less. Another requirement imposed by Part 15 is that 2.4 GHz devices must use spread spectrum transmission and modulation techniques. *Modulation* is a method of varying the radio signal in a way that information is conveyed to the receiver. *Spread spectrum* is a way of spreading the radio signal over a wider frequency band than the minimum required for transmission of data at a specific rate. This technique sacrifices bandwidth usage efficiency for decreased interference. The two spread spectrum techniques defined by Part 15 are *direct sequence spread spectrum* (DSSS) and *frequency-hopping spread spectrum* (FHSS). Modulation techniques are key differences between IEEE 802.11, IEEE 802.11a, IEEE 802.11b, and IEEE 802.11g. The combination of frequency, bandwidth, and modulation limited the original IEEE 802.11 to 2 Mbps. The modification of these characteristics has increased its descendant's speed to 54 Mbps. It's important to note that although PHY-level radio characteristics like frequency and power are regulated by the FCC's Part 15 rules in North America, regulatory agencies in other parts of the world have different requirements. Frequency and transmit power requirements are shown in Table 2-1. Modulation and spread spectrum will be covered in detail in Chapter 6.

Table 2-1

WLAN frequencies and power

Standard	Frequency	Maximum Transmit Power
802.11 802.11b 802.11g	North America 2.4–2.4835 GHz	1,000 mW
	Europe 2.4–2.4835 GHz	100 mW
	Japan 2.471–2.497 GHz	10 mW/MHz
	Spain 2.445–2.475 GHz	100 mW
	France 2.4465–2.4835 GHz	100 mW
802.11a	North America 5.15–5.25 GHz	40 mW
	5.25–5.35 GHz	200 mW
	5.47–5.725 GHz (proposed)	
	5.725–5.825 GHz	800 mW

Table 2-1 *cont.*

WLAN
frequencies
and power

Standard	Frequency	Maximum Transmit Power
High-performance radio local area network2 (HIPERLAN/2)	Europe 5.25–5.35 GHz 5.47–5.725 GHz	200 mW (indoor) 1 W (indoor/outdoor)
HiSWAN	Japan 5.15–5.35 GHz	200 mW (indoor/outdoor)

IEEE 802.11—A Mobile Network Architecture

IEEE 802.11 defines a hierarchical network architecture that enables WLAN equipment to be configured in a variety of ways. End-points are called *stations* and are connected to the WLAN through a wireless *network interface card* (NIC). The simplest configuration is called an *independent basic service set* (IBSS). An IBSS is a self-contained network that only supports direct communication between stations that are members of that network. A typical example would be three people with laptops sitting in a park and establishing an ad hoc peer-to-peer network. See Figure 2-1. Each station within an IBSS may directly communicate with one or more stations within the group. There is no master station or centralized coordination of the group. All communication is directly between stations.

Figure 2-1
An IBSS

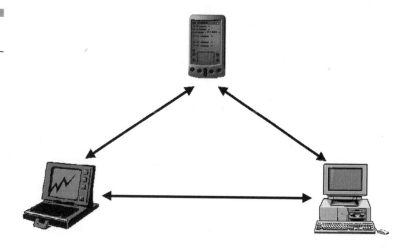

The second type of configuration is called an infrastructure *basic service set* (BSS). A BSS adds centralized distribution, buffering, and gateway functionality through a device called an *access point* (AP). All communications in an infrastructure BSS pass through the AP (see Figure 2-2). Because an AP relays messages from station to station, it effectively doubles the area covered. Its buffering capabilities allow power management for battery-powered devices, and its gateway functionality provides connectivity to a wired LAN or external network. An AP providing a gateway to an external network is also called a *portal*. The centralized management provided by an AP enables vendors to package other capabilities such as firewalls and print spoolers in the same equipment. Although an AP provides mobility within the area covered by its radio signal, larger areas may require multiple APs. Network configurations with multiple APs, each controlling their own BSS, make up the next level in the hierarchy.

Figure 2-2
An infrastructure basic service set (BSS)

A configuration with multiple infrastructure BSSs functioning as a single WLAN is called an *extended service set* (ESS) (see Figure 2-3). An ESS enables each AP within its infrastructure BSS to determine if messages are to be sent to a station within its BSS, passed to another AP's BSS, or transmitted to an external LAN or network.

This mechanism in IEEE 802.11 is called the *distribution system* (DS). The DS is a logical concept. It is independent of the media connecting an AP to another element of the DS. This enables an AP to communicate to the stations via radio while having a choice of media (radio, wired LAN, and so on) for communication between APs or external networks. ESSs appear to be a single network to external entities, masking the mobility implied by a station's capability to move seamlessly between APs.

Figure 2-3
An extended
service set (ESS)

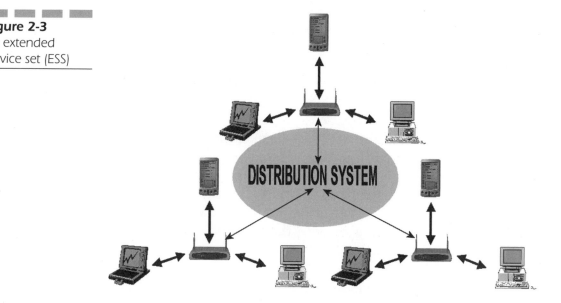

DISTRIBUTION SYSTEM

The capability to roam within an ESS introduced a variety of requirements related to *Internet Protocol* (IP) addressing and network configuration. Although IEEE 802.11 defined the architecture of an ESS and the basic services that enable the delivery of data over the DS, it was not specific on how APs and portals should be implemented. This gave vendors a great deal of flexibility for AP implementation. The standardization of interaccess point communications is one of the weaker areas of the IEEE 802.11 standard. This weakness combined with the flexibility of the network architecture led to incompatibility between APs from different vendors. Recognizing this, the IEEE 802.11 Committee formed Task Group *f* (IEEE 802.11f) to address this problem.

Another area related to network configuration that IEEE 802.11 did not address was the capability to have a MAC bridging function between two LANs. Task Group *c* (IEEE 802.11c) was formed to address the bridging requirements of WLANs. Upon approval, the recommendations of the IEEE 802.11c Task Group were incorporated into the IEEE MAC bridging standard, IEEE 802.1d. IEEE 802.11c does not exist as an independent specification.

IEEE 802.11f—The Interaccess Point Protocol (IAPP)

The need for an IAPP has been recognized for quite a while. As early as 1996, Lucent Technologies, Aironet Corporation (now part of Cisco), and Digital Ocean started work on a cooperative IAPP. Although various proposals were discussed among major vendors, they were not widely implemented. Market dynamics were such that early IEEE 802.11 systems tended to be small in scope and used a single vendor's hardware; as a result, standardized roaming was not a pressing issue. With the advent of IEEE 802.11b in 1999, accelerating deployments, and the growing proliferation of IEEE 802.11 hardware vendors, this changed. The IEEE 802.11 Working Group assigned the problem of interaccess point communication to a Study Group, and in March 2000, a project authorization was approved forming IEEE 802.11f. The purpose of IEEE 802.11f is to standardize the way a DS is created and managed. It specifies the information and practices required for managing the ESS DS. This includes functions such as new AP registration, AP addressing, and the maintenance of roaming station information. Although still in the draft stage as of this writing, current proposals include the use of standard protocols such as the *Service Location Protocol* (SLP) for registration services. Approval is expected in 2002.

The IEEE 802.11a Standard

The IEEE 802.11a standard was published in 1999 to address the need for faster data transmission rates. It accomplished this by using the higher-frequency 5 GHz band and employing a radio transmission technique called *orthogonal frequency division multiplexing* (OFDM). OFDM is emerging as a key technology in several wireless communications protocols. IEEE 802.11a supports data rates of 6, 9, 12, 18, 24, 36, 48, and 54 Mbps and is intended to operate in the United States using the FCC's *Unlicensed National Information Infrastructure* (U-NII) band. This band is comprised of 300 MHz of radio spectrum between 5.15 and 5.35 GHz and 5.725 and 5.825 GHz.

IEEE 802.11a and HIPERLAN1/2

The 5 GHz band is also used by the HIPERLAN/1 and HIPERLAN/2 standards. HIPERLAN/1 was the European community's first standard for WLANs. It supported speeds up to 24 Mbps. Primarily designed for very simple network architectures, it has since been superceded by the far more robust HIPERLAN/2. HIPERLAN/2 is the most recent WLAN standard published by the *Broadband Radio Area Network* (BRAN) project of the *European Telecommunications Standards Institute* (ETSI). Like IEEE 802.11a, HIPERLAN/2 supports speeds up to 54 Mbps. It is intended to work worldwide as a 5 GHz WLAN standard. The PHY layer is almost identical to IEEE 802.11a. They mainly differ at the MAC layer and in the regulation of the 5 GHz frequency worldwide. IEEE 802.11a provides a MAC that functions as a wireless version of the IEEE 802.3 MAC. HIPERLAN/2, on the other hand, is designed to support time-critical applications and has defined interfaces to *third-generation* (3G) networks, *Asynchronous Transfer Mode* (ATM) networks, and Firewire (IEEE-1394) networks as well. In addition, ETSI specified *transmission power control* (TPC) and *dynamic frequency selection* (DFS) for better spectrum utilization and reduced interference. Because one of the goals of any IEEE standard is to consider compatibility with other standards, this led the IEEE 802 Committee to form Task Group *h* (IEEE 802.11h) and 5GSG to make recommendations regarding IEEE 802.11a and its relationship to HIPERLAN/2.

IEEE 802.11h, 5GSG, and IEEE 802.11d— International Operation

The IEEE 802.11h Task Group is making recommendations for the reconciliation of TPC and DFS capabilities specified in HIPERLAN/2 with those specified by IEEE 802.11a. This will require enhancements to the PHY level of IEEE 802.11a to support the 5.47 to 5.725 MHz frequency bands as well as changes at the MAC layer protocol to support TPC and DFS. The IEEE 802.11h recommendations are in letter ballot with final approval expected in 2002.

5GSG is an interesting initiative as it is a cooperative venture with the representatives of ETSI BRAN and Japan's Promotion Council for *Mobile Multimedia Access Communication* (MMAC). The goal is to reconcile the major standards for 5 GHz WLANs to arrive at a single global standard. These standards are IEEE 802.11a, HIPERLAN/2, and Japan's HiSWAN and *carrier sense multiple access* (CSMA). Task Group IEEE 802.11h is a specific example of this cooperation.

The IEEE 802.11 specification accommodates the bandwidth regulations in North America (FCC), Europe (ETSI), Japan, Spain, and France. Other regulatory domains have different rules on FHSS and DSSS bandwidth and channel usage. With the growing popularity of IEEE 802.11, Task Group *d* (IEEE 802.11d) was formed to allow IEEE 802.11 operation in additional regions such as Korea, Israel, Mexico, and Belgium. The IEEE 802.11d specification was approved in June 2001.

IEEE 802.11a and 802.11b Coexistence

The use of the 5 GHz frequency band makes IEEE 802.11a fundamentally different from IEEE 802.11 and its 2.4 GHz descendents IEEE 802.11b and IEEE 802.11g. Although the IEEE 802.11b market will continue to grow quickly for several years, the 5 GHz standards, IEEE 802.11a and HIPERLAN/2, are increasingly seen as long-term commercial successors to IEEE 802.11b. IEEE 802.11b has gained commercial traction more quickly than IEEE 802.11a for several reasons. It uses the same 2.4 GHz frequency as the original

IEEE 802.11, and component designs were able to leverage previous work in the 2.4 GHz band. The use of a lower frequency makes IEEE 802.11b less susceptible to radio signal degradation and enables it to cover a larger area. Its modulation schemes are easier to implement and it consumes less power. Although it is easier to bring to market, the 2.4 GHz frequency is physically limited in the amount of data it can carry relative to a 5 GHz frequency. The 5 GHz band not only allows faster data rates, but it is also less densely populated with interfering equipment (such as cordless phones and Bluetooth devices) than the 2.4 GHz band.

To overcome incompatibility at the radio modem level and ease migration issues, vendors are creating multimode chip and equipment designs that allow the use of both the 5 and 2.4 GHz bands simultaneously. Chip manufacturers like IceFyre are introducing sophisticated IEEE 802.11a technologies that reduce power consumption. The second half of 2001 saw an acceleration of IEEE 802.11a product releases as vendors overcame cost and design barriers. The demand for bandwidth-intensive applications (such as streaming media) is increasing, and standards such as IEEE 802.11e that support those applications are emerging. This combination will drive the demand for IEEE 802.11a equipment in 2002 and 2003.

The IEEE 802.11b Task Group

The IEEE 802.11b standard was also published in 1999 to address the need for data transmission speeds that are higher than those of IEEE 802.11. Utilizing the same 2.4 GHz frequency band, it is a natural extension of the original IEEE 802.11 standard. It is designed to be backward compatible with both frequency-hopping and direct sequence IEEE 802.11 systems while allowing operation at speeds of 1, 2, 5.5, or 11 Mbps. It accomplishes this through spread spectrum modulation schemes that result in higher data rates while providing mechanisms within the protocol for the discovery and negotiation of a common transmission scheme between two devices. The primary DSSS scheme supporting these higher data rates is called *complementary code keying* (CCK). Another optional scheme called *packet binary convolutional coding* (PBCC) was also included.

As we will see, the inclusion of PBCC as an optional scheme in IEEE 802.11b has important implications for IEEE 802.11g, the new proposed standard for 22 Mbps operation. IEEE 802.11b provides the capability for 5.5 and 11 Mbps direct sequence systems to interoperate and coexist with legacy IEEE 802.11 systems. It also has the capability to operate in areas where frequency-hopping regulations may be different than those specified for North America by the FCC. It accomplishes this through a capability called *channel agility*, which allows the negotiation of the modulation scheme and transmission rate.

The IEEE 802.11g Task Group—And in This Corner . . .

In the search for faster data rates in the 2.4 GHz band, IEEE 802.11 (2 Mbps) gave way to IEEE 802.11b (11 Mbps). The installed base of 2.4 GHz equipment and the market's constant demand for higher data rates led to the formation of Study Group HRb (high-rate IEEE 802.11b) in March 2000. Their work led to the formation of the IEEE 802.11g Task Group in September 2000. The IEEE 802.11g Task Group was formed to produce a standard for 22 Mbps data rates in the 2.4 GHz band. Without increased bandwidth or a different frequency, this can be achieved through the use of more sophisticated modulation techniques. The IEEE 802.11g Task Group has become the battleground for several competing modulation schemes. Intersil and Texas Instruments, two of the largest suppliers of IEEE 802.11 chips and components, are the main combatants in the IEEE 802.11g arena.

The seeds for this battle were sown when PBCC was included in the IEEE 802.11b specification as an option to CCK. PBCC was pioneered by Alantro Communications Corporation. Seeking to expand their share of the growing IEEE 802.11 market, Texas Instruments acquired Alantro in June 2000 for approximately $300,000,000. Texas Instruments has continued to develop PBCC, has achieved rates of 22 Mbps, and is the dominant force in the PBCC chip market. It is understandable, therefore, that Texas Instruments has proposed PBCC as a way of achieving the 22 Mbps data rates required by IEEE 802.11g. Texas Instruments argues that PBCC is a natural extension to IEEE 802.11b and requires less protocol overhead, resulting in greater data throughput. Texas Instruments has offered to license its PBCC technology for free in order to alleviate concerns that it would have a technological advantage over other industry participants. If PBCC is adopted to the exclusion of other techniques, Texas Instruments would still maintain a significant time-to-market advantage because of its ongoing work with PBCC.

In the other corner, Intersil, as a part of Harris Semiconductor, joined forces with Lucent in 1999 to establish CCK as the primary modulation scheme for direct sequence IEEE 802.11b LANs. Intersil became a dominant force in CCK-based IEEE 802.11b chips. Intersil has proposed OFDM as a way of achieving the 22 Mbps data rates in the 2.4 GHz band. Intersil argues that OFDM is a widely understood and implemented technology that has also been adopted as the modulation technique for IEEE 802.11a WLAN and IEEE 802.16 *metropolitan area network* (MAN) standards. It has proposed extensions to the IEEE 802.11b protocol headers that would provide additional mandatory data rates of up to 24 Mbps. These would be backward compatible with IEEE 802.11b systems, but Texas Instruments argues that data throughput would be less because of the additional protocol overhead. Intersil's proposal also provides optional rates up to 54 Mbps that are not backward compatible with IEEE 802.11b systems, but uses OFDM radio transmission techniques that are compatible with IEEE 802.11a.

This debate has been further clouded because of the current FCC regulations governing the 2.4 GHz band. There is an ongoing debate about whether OFDM technologies are considered to be DSSS techniques under the current regulations. In May 2001, the FCC issued a Notice of Proposed Rulemaking that was designed to relax the rules regarding DSSS and FHSS systems in the 2.4 GHz band. In order to reduce interference, it allowed new adaptive frequency-hopping algorithms. In addition, it proposed allowing OFDM systems to operate in the band and granted an immediate waiver for limited-power wideband OFDM (W-OFDM) systems manufactured by Wi-LAN. This proposed rulemaking is in the public comments phase, but it appears to have support. This rulemaking would have to be adopted for OFDM IEEE 802.11g systems to be deployed.

The commercial implication of adopting one of these proposals over the other has caused a lengthy debate over the procedures used to achieve a 75 percent majority vote. The IEEE 802.11g Task Group has spent as much time debating parliamentary voting procedures as they spent debating the technical merits of the proposals. This has resulted in an inordinate delay in the approval of a standard and opened a window of opportunity for the introduction of IEEE 802.11a equipment. In mid-November, the IEEE 802.11g Task Group finally approved a draft standard for 22 Mbps operation in the 2.4 GHz frequency band. It is a compromise of sorts and the marketplace will decide if there is a clear winner. The IEEE 802.11g draft standard incorporates two mandatory modes: an OFDM mode to provide compatibility with 802.11a for speeds up to 54 Mbps and an IEEE 802.11b CCK mode to provide backward compatibility with IEEE 802.11b systems. Two optional modes are also supported for data rates up to 24 Mbps: PBCC-22 and CCK-OFDM. The support of both IEEE 802.11b and IEEE 802.11a provides an opportunity to produce dual-band equipment spanning the 2.4 GHz and 5 GHz worlds. This may extend the life of 2.4 GHz systems and fill the need for higher-speed systems until the maturity of 5 GHz WLANs.

The IEEE 802.11e Task Group—QoS

The IEEE 802.11 specification addressed the basic transport of LAN data over a wireless medium. It did not fully address mobility, security, and other higher-level functions required to support sophisticated applications such as streaming media and voice. These applications have unique requirements for time-sensitive packet delivery as opposed to the less stringent requirements imposed by applications such as file transfer. Wired IEEE 802 networks were designed to handle bursts of traffic in the most efficient fashion. The reliability and predictability of packet delivery were traded for efficiency and ease of implementation. Streaming media applications require steady, predictable data packet delivery and do not function well on basic IEEE 802 LANs. IEEE 802.11 networks inherited this characteristic from their wired brethren. The basic transport mechanism enables packets to travel different routes to their final destination. This means that they may have different arrival rates. Although it may not affect the end result of a file transfer, packets that are delayed or arrive at irregular intervals may make a voice conversation or media stream unintelligible. Packet delay and jitter (irregular arrival times) are unacceptable for any sort of audio, video, or steaming media.

The requirement for a reliable, predictable data stream is typically resolved by providing the ability to classify traffic and prioritize time-sensitive classes of traffic. This is generally referred to as QoS. Traffic classification can be provided in different ways at different layers of an IEEE 802 protocol. In the wired world, this capability is provided by data link (*Open Systems Interconnection* [OSI] layer 2) mechanisms such as the prioritized Ethernet capabilities of IEEE 802.1p. IEEE 802.1p provides eight classes of traffic. There are OSI layer 3 protocols like *Differentiated Services* (Diffserv), higher-layer protocols like the *Resource Reservation Protocol* (RSVP), and even protocol-independent schemes like *Multiprotocol Label Switching* (MPLS).

The unique characteristics of data transport over a WLAN require a QoS mechanism specific to IEEE 802.11. This is the charter of Task Group *e* (IEEE 802.11e). The initial charter of IEEE 802.11e included security as well as QoS capabilities. The project authorization for IEEE 802.11e was first approved in March 2000 and the functional requirements were approved in July of that same year. By early 2001, it became apparent that QoS and security were distinct capabilities and required separate Task Groups. This resulted in the IEEE 802.11e Task Group narrowing its charter to focus entirely on QoS. In March 2001, Task Group *i* (IEEE 802.11i) was formed to specifically address the issue of security.

The 802.11e Task Group has prepared a draft specification that is being circulated for comment. It includes enhancements that allow more effective channel management, resulting in increased throughput and reduced power requirements. These enhancements include *forward error correction* (FEC) algorithms, the ability to selectively retransmit data, and improvements to the polling algorithms used by APs. In addition, each stream of data is able to negotiate the most effective means of transmission. Final approval is expected in late 2001 or early 2002.

IEEE 802.11i Task Group—Security, Encryption, and Authentication

One of the greatest inhibitors to the adoption of IEEE 802.11 LANs has been the security problems introduced by broadcasting data over a wireless medium. Wired LANs require physical access to the wire. Wireless data is accessible to anyone within range of the antenna and tuned to that frequency. Once access to any transmission medium is gained, the network is subject to a variety of security problems, including access to private data and the injection of destructive data into the network. The only way to prevent someone with access to the transmission media (wired or wireless) is through sophisticated security capabilities such as authentication, encryption, and network firewalls.

The greatest strengths of IEEE 802.11 (the availability of low-cost standardized equipment and open access to the transmission medium) are its greatest weaknesses in the area of security. The designers of IEEE 802.11 did not ignore this fact. They incorporated a scheme called the *Wired Equivalency Protocol* (WEP) into IEEE 802.11. It was intended to provide a level of security for a WLAN that was equivalent to a wired LAN where physical access was prevented. It used an encryption scheme called RC4 with 64-bit keys and provided very basic facilities for security key management. These capabilities proved inadequate as security technology advanced, and in early 2001, researchers at the University of California at Berkeley published a paper that showed how to compromise the WEP security scheme. Armed with low-cost equipment, instructions on how to breach WEP, and easy access to IEEE 802.11 radio signals, it became relatively easy to breach a WLAN. A practice called *war driving* started to appear. War driving is the wireless equivalent to *war dialing*. During the 1970s and 1980s, war dialing was the practice of systematically dialing a series of phone numbers with a computer until access to another computer was gained. In war driving, a hacker drives around in a vehicle with an IEEE 802.11-enabled computer until an IEEE 802.11 LAN signal is encountered. It is then possible to use the procedures described by the Berkeley researchers to compromise the IEEE 802.11 LAN. Several strategies have been proposed and implemented outside of the IEEE 802.11 specification to overcome this deficiency, but because they are outside of the standards process, they introduce problems of interoperability, complexity, and backward compatibility.

In order to address the security issue within the IEEE 802.11 framework, Task Group *i* (IEEE 802.11i) was formed to standardize the security mechanisms for WLANs. Formed in March 2001, the IEEE 802.11i Task Group has been investigating authentication schemes like Kerberos and *Remote Access Dial-In User Service* (RADIUS). There has been a proposal to enhance WEP (called *WEP2*) where the encryption key is increased to 128 bits. Unfortunately, it uses the same RC4 encryption scheme and WEP's key management; therefore, it is still susceptible to the Berkley researchers' techniques. Also under consideration is the *Advanced Encryption Standard* (AES). This is a proposed U.S. federal standard specifying a cryptographic algorithm for use by U.S. government organizations to protect sensitive (unclassified) information.

Incorporating these very complex technologies into IEEE 802.11 while balancing the need for interoperability, simplicity, backward compatibility, and speedy resolution is a difficult task. The IEEE 802.11i Committee has prepared a draft that is now being circulated for comment. Given the complexity of the problem, a published standard before mid-2002 is unlikely. Fortunately, there are a number of strategies that minimize the security risks of deploying an IEEE 802.11 LAN. There is a detailed discussion of WLAN security and these strategies in Chaper 8.

IEEE 802.11 Wireless Relatives

As introduced in Chapter 1, "802.11—Bridging the Gap," the *Institute of Electrical and Electronics Engineers* (IEEE) 802.11 family is one of several wireless data communications technologies. As we will see, the period from 1997 through 1999 was a seminal period for the emergence of wireless networking technologies and standards other than IEEE 802.11. Encompassing both light- and radio-based media, these methods are conveniently grouped by the distance that they generally serve: short range, local, regional, and global. Although Vint Cerf has a project underway with NASA's Jet Propulsion Laboratory to create an interplanetary Internet, we'll limit ourselves to terrestrial distances. This chapter presents the technologies that deliver wireless communications within each of these domains.

Wireless Personal Area Networks (WPANs)

Bluetooth

Bluetooth has emerged as the dominant wireless networking technology for distances less than 10 meters. Despite the economic woes of 2001, Bluetooth chipset shipments have continued to grow rapidly and are expected to be more than 45,000,000 units in 2002. Although commercial availability of Bluetooth-enabled equipment has lagged behind IEEE 802.11b, with chipset prices projected to approach $5 in late 2002, Bluetooth will soon become a standard feature of consumer electronics, cellular phones, computer peripherals, and mobile computing appliances.

The L.M. Ericsson company created Bluetooth in 1994 and named it after the tenth-century Danish king Harald Blaatlund "Bluetooth" II. Bluetooth started to gain commercial momentum with the formation of the Bluetooth *Special Interest Group* (SIG) in 1998. Charter members of the Bluetooth SIG were Ericsson, Intel, IBM, Nokia, and Toshiba. They joined together to create and promote an open standard for short-range wireless networking. Although the specification for Bluetooth is license and royalty free, membership in the

SIG is required for use and access. The Bluetooth SIG has three levels of membership: promoters, associates, and adopters. In addition to the charter members, promoters now include Microsoft, 3Com, Motorola, and Lucent. There are over 2,000 adopter and associate companies. Adopter companies do not pay a membership fee. They also get a royalty-free license and the right to use the Bluetooth brand. Associate members pay membership fees of $7,000 or $40,000 a year depending on whether annual corporate revenues are greater than $100 million. Associates have the same rights as adopters, but they are also given early access to specifications and participation in the organization's Working Groups.

The latest specification is version 1.1, which was developed in February 2001. Version 2.0 is under discussion. One of the main items under discussion is the speed of the 2.0 specification. There are two main approaches being considered. The first would increase the speed to 2 Mbps, but retain lower costs and backward compatibility with existing systems. The second proposal would increase speeds to 10 Mbps or more, but come with increased cost and a lack of backward compatibility. A parallel effort to create a high-speed *wireless personal area network* (WPAN) is underway in the IEEE 802.15.3 Working Group.

The recent growth of Bluetooth has paralleled and complemented the growth in IEEE 802.11b. As microprocessors found their way into more devices (it's estimated that 90 percent of all microprocessors are in devices other than personal computers), these embedded systems required a small-form-factor, low-power, short-range means of communication. The network architecture supported by Bluetooth is also optimized for a large number of devices communicating as equals in a peer-to-peer network. In a Bluetooth network, devices discover one another as they move into each other's range. Connections between devices are established in a master/slave relationship, with a master having up to seven active slaves. A master may suspend its connection to a slave with a technique called *parking* and establish a connection to a new active device. The master may have a total of 256 connections with any 7 active at one time. This set of master/slave relationships is called a *piconet*. A device can be a master of a piconet, but also simultaneously be a slave to another device in another piconet. This series of overlapping piconets is called a *scatternet.* This network topology supports very dense concentrations of devices in a small area and enables devices to quickly establish links to each other as they move in and out of range. Piconets and scatternets are shown in Figure 3-1.

The protocol layers of Bluetooth map to the physical and data link layers of the *Open Systems Interconnection* (OSI) model much like the IEEE 802.11 series of standards. Bluetooth's physical layer is called the *radio layer*, and its data link layer is subdivided into *baseband* (for accessing the radio) and *Logical Link Control* (LLC) functions. In addition, it specifies a control interface to the host controller and special LLC for audio processing. There are also several upper-layer specifications. These include a protocol for locating available Bluetooth services called the *Service Discovery Protocol* (SDP), an RS-232 serial port emulation protocol called *RFCOMM*, a *Telephony Control Services* (TCS) protocol, and an infrared emulation protocol called *IrOBEX*.

Bluetooth employs *frequency-hopping spread spectrum* (FHSS) techniques and uses frequencies in the 2.4 GHz radio band like IEEE 802.11 and 802.11b. A Bluetooth radio in asymmetric mode supports a maximum data rate of 721 Kbps in one direction and 57.6 Kbps in other. It supports 432.6 Kbps in each direction in symmetric

Figure 3-1
Bluetooth network topology

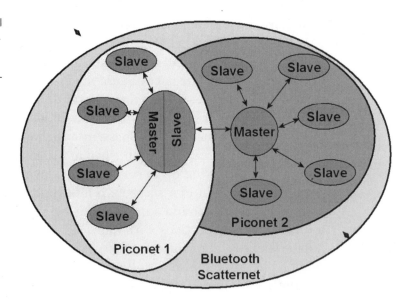

mode. Its transfer protocol is designed to handle both voice and data and supports *forward error correction* (FEC) and encryption. Frequency hops are done after each packet is sent or received. Bluetooth uses 79 hopping channels, which are 1 MHz wide, and uses a fast hopping scheme (1,600 hops/sec). In contrast, the IEEE 802.11 FHSS scheme specifies hopping rates as low as 2.5 hops/sec. The *direct sequence spread spectrum* (DSSS) scheme of IEEE 802.11b uses 11 overlapping 22 MHz channels.

The fact that IEEE 802.11 and IEEE 802.11b products operate in the same frequency range as Bluetooth has led to concerns about interference as Bluetooth products are deployed in quantities. Interference between Bluetooth at 2.4 GHz and IEEE 802.11a at 5 GHz is not an issue as they utilize different frequencies. Generally speaking, interference is nominal if the distance between Bluetooth and IEEE 802.11 or 802.11b devices is more than 3 meters. As distances between devices lessen, there is the possibility of increased interference that affects reliability and speed. This makes interference especially troublesome when an IEEE 802.11b and Bluetooth radios are colocated in the same device. An example of this would be a computer simultaneously using 802.11b for *local area network* (LAN) connectivity and Bluetooth for a wireless headset. In general, because Bluetooth uses a greater number of smaller rapidly changing channels and smaller packets, it is less susceptible to interference than IEEE 802.11 and IEEE 802.11b systems.

The one area where Bluetooth has a low tolerance for interference is in its handling of synchronous data such as voice. Packet delivery of synchronous data is very sensitive to variations in the rate at which data is received, and even low levels of interference can affect quality. Because of the differential in hopping rates between 802.11 and Bluetooth, interference resulting in performance degradation is likely for older IEEE 802.11 systems using FHSS. Although 802.11b uses DSSS, it still shares the 2.4 GHz band and interference is possible if a Bluetooth device at a high enough power level hops into the 22 MHz band it is using. The factors affecting interference levels are dynamic in nature, making it a difficult problem to quantify. These factors include the number of colocated devices, power levels, the width of the respective radio bands being used, and the duration of packet transmission.

Fortunately, regulatory agencies, vendors, and standards bodies are addressing the issue of 2.4 GHz interference. The first notable proposal is an amendment to *Federal Communications Commission* (FCC) Part 15.247. This is the set of regulations that specify spread spectrum methods in the 2.4 GHz radio band. The amendment will enable adaptive-hopping algorithms to be used in the 2.4 GHz band. Adaptive-hopping algorithms enable devices to change their hopping sequence based on previously encountered interference. Vendors are starting to release Bluetooth products that use adaptive frequency-hopping techniques. Bluetooth chipset maker Bandspeed Inc., in conjunction with Open Interface North America, has announced a chipset that employs adaptive frequency hopping and is targeted for use in IEEE 802.11b environments. Mobilian has announced a chipset that handles Bluetooth and IEEE 802.11b simultaneously. Access to both the radio and baseband processing allows chip-level management of the signals for reduced interference.

Organizations such as the IEEE and the *National Institute of Standards and Technologies* (NIST) are also actively involved in addressing the issue of Bluetooth and IEEE 802 *wireless LAN* (WLAN) coexistence. Task Group 2 of the IEEE 802.15 Working Group on WPANs is devoted to studying 802.11b/Bluetooth coexistence issues. In April 2001, NIST, Mobilian, and Symbol Technologies submitted a proposal to the IEEE 802.15.2 Working Group for collaborative coexistence between Bluetooth and IEEE 802.11b. *Collaborative coexistence* means that a Bluetooth receiver is deployed as part of the IEEE 802.11b system. This enables the IEEE 802.11b system to discover information about the Bluetooth hopping pattern and avoid interference.

As evidence that Bluetooth and IEEE 802.11b can coexist, *United Parcel Service* (UPS) has announced a $100,000,000 deployment of Bluetooth and 802.11b devices for its package-handling operations. The application includes Bluetooth-enabled package scanners that are worn on the hand as rings and IEEE 802.11b terminals running Microsoft Windows CE. Up to 15,000 802.11b *access points* (APs) and 50,000 terminals may be deployed. Symbol Technologies provides the scanners and APs, and Motorola supplies the terminals. The cordless scanners eliminate the safety hazards posed by cords around conveyer belts and increase productivity by freeing workers' hands for other tasks.

It would seem that Bluetooth and 802.11 have considerable overlap, but the design priorities of the two technologies have resulted in a complementary relationship. Bluetooth trades speed and range for lower cost and power requirements—two primary considerations for a network of many small devices located relatively close to one another. The IEEE 802.11 series of standards is designed as a wireless alternative to wired LANs and is optimized for faster speeds, greater range, and easier connectivity to the existing wired LAN infrastructure. Although Bluetooth has the capability to operate at higher power levels and ranges greater than 10 meters, and although IEEE 802.11b components reduce power consumption and cost, each is still firmly centered in its respective market segment.

IEEE 802.15

In early 1998, around the same time that Bluetooth was starting to gain commercial recognition, the IEEE 802.11 Working Group formed a Study Group to examine low-power, short-range wireless networks. As it became apparent that the wireless networking need of embedded systems was a fundamentally different problem than WLANs, the IEEE 802 Committee created the IEEE 802.15 Working Group to address standards for WPANs. There are four active Task Groups in the IEEE 802.15 Working Group (numbered 1 through 4).

Task Group 802.15.1 has the job of delivering an IEEE standard for low-speed, low-cost WPANs. In mid-1999, the IEEE 802.15.1 Task Group decided to use the Bluetooth specification as the basis for the IEEE WPAN standard. It was conditionally approved as a new standard in March of 2002

Task Group 802.15.2 is developing recommended practices on how IEEE 802.11 WLANs and IEEE 802.15 WPANs can coexist in the 2.4 GHz band. Because 802.15.1 is based on Bluetooth, the IEEE 802.15.2 Task Group is in large part tackling the problem of interference between Bluetooth and IEEE 802.11, as discussed previously. It is formulating recommendations that include adaptive power control, adaptive frequency hopping, and collaborative approaches to reducing interference.

Task Group 802.15.3 has the job of delivering an IEEE standard for higher-speed WPANs. These WPAN networks will support speeds of 10 to 55 Mbps at distances that are less than 10 meters. Intended for multimedia and digital media applications, the standard also includes *quality of service* (QoS) and security features. One drawback to the higher speeds is that the supporting radio technology may not be backward compatible with Bluetooth 1.1. The IEEE 802.15.3 standard is scheduled to be completed in late 2002.

At the other end of the speed spectrum from IEEE 802.15.3, Task Group 802.15.4 is preparing a standard for simple, low-cost, low-speed WPANs. One of the major design goals is to extend battery life (by months or years). These WPANs would be used in applications such as remote sensing, industrial control, home automation, medical devices, toys, smart cards, and other small-form-factor, low-throughput devices. Data rates range from 2 to 200 Kbps and use DSSS modulation. The proposed *radio frequencies* (RFs) are 2.4 GHz and 915 MHz for North America and 868 MHz for Europe. The 802.15.4 standard is scheduled to go to letter ballot in late 2002.

Digital Enhanced Cordless Telecommunications (DECT)

The *Digital Enhanced Cordless Telecommunications* (DECT) standard has been in existence since 1992. Maintained by the *European Telecommunications Standards Institute* (ETSI), it is designed to provide wireless communications for telecommunications equipment such as cordless phones and *private branch exchanges* (PBXs). It operates in the 1.9 GHz radio band and employs *Time Division Multiple Access* (TDMA) technology, making it well suited for voice applications. It is promoted by the DECT Forum, an industry trade group. DECT has operated at speeds of 2 Mbps for several years, but a proposal by ETSI to increase data rates to 20 Mbps and its range of 17 kilometers would seem to make DECT a potential competitor for IEEE 802.11 and Bluetooth systems.

Several factors are inhibiting the adoption of DECT as a general-purpose, short-range wireless standard. First and foremost, the difference in licensing the 1.9 GHz band between Europe and North

America is a barrier to worldwide deployment. Secondly, as QoS methods are deployed, voice is transmitted more often over *Internet Protocol* (IP) networks. The blurring of voice and data networking reduces the need for voice-only networks. Lastly, the rapid adoption rates of Bluetooth and IEEE 802.11 systems worldwide are narrowing the market space available to DECT. Although it is still valuable as a cordless telephony standard, DECT has significant barriers to overcome before it can be considered a general-purpose networking standard.

Infrared Wireless Networking

The wireless networking methods that we have discussed so far have been radio based, but there are light-based methods as well. In the WPAN arena, these are implemented with infrared light. Infrared light is an invisible type of electromagnetic radiation at the low end of the electromagnetic spectrum. If you have ever changed the station on a television with a remote control, chances are you've used infrared networking. In general, infrared-based networking methods can be divided into two categories: direct beam and diffused beam. *Direct beam infrared*, as the name implies, is a point-to-point communication method requiring a clear line of sight between the two devices. *Diffused beam infrared* uses the reflected energy of the light beam to transmit data. Diffused beam systems do not require line of sight, but the use of reflected energy makes data rates slower.

The advantages of infrared systems are that they are for the most part unregulated, inexpensive, and immune to interference from higher-frequency communication systems such as radio. However, there are also several limitations. Direct beam systems cannot penetrate solid objects, their range is usually just a few meters, and they are unsuitable for outdoor use due to interference from environmental conditions like rain and fog. As you may recall, the original 802.11 standard contains an infrared *physical* (PHY) layer that has never been widely deployed for WLANs due in large part to these limitations. On the other hand, infrared systems are arguably the most successful point-to-point, short-range networking technology to date.

They are already deployed in millions of laptops, *personal digital assistants* (PDAs), cell phones, consumer electronics, and other devices worldwide. The main group involved in creating and promoting today's infrared networking systems is the *Infrared Data Association* (IrDA).

IrDA has created a set of short-range, point-to-point infrared networking standards. There are currently two primary standards in the IrDA family: IrDA-Control and IrDA-Data. IrDA-Control is a low-speed protocol optimized for use with wireless control devices such as computer mice, joysticks, gamepads, remote controls, and other devices. The use of the IrDA-Control standard has brought forth applications such as the universal remote control. Like IEEE 802 systems, IrDA-Control has PHY, *Medium Access Control* (MAC), and LLC protocol elements. The IrDA-Control protocol operates at 75 Kbps and has very low power requirements. A host may communicate with up to eight peripherals simultaneously; the link control also has reliable data delivery mechanisms such as sequencing and retransmission.

IrDA-Data is a wireless data networking standard like Bluetooth or 802.15. It is a bidirectional, point-to-point protocol designed to operate at distances that are less than 1 meter and at speeds from 9.6 Kbps to 16 Mbps. Range can be extended to 2 meters, but at the cost of greater power consumption. A low-power version sacrifices range (operation at 20 to 30 cm), but it reduces power consumption by a factor of 10. The IrDA protocol stack has three mandatory layers: PHY, the *Infrared Link Access Protocol* (IrLAP), and the *Infrared Link Management Protocol* (IrLMP). The IrLAP defines point-to-point device connection methods, including the discovery of devices. The IrLMP allows multiple logical connections and manages IrDA-Data services through the Information Access Service.

There are also several optional protocols enabling functions like *infrared serial/parallel port emulation* (IrCOMM), *infrared digital image transfer* (IrTran-P), *infrared local area network connectivity* (IrLAN), and *infrared mobile communications* (IrMC). These protocols extend IrDA-Data for use in a specific application. For example, IrMC specifies the exchange of audio streams and data used in mobile handset applications (such as phone books, text messages, contacts, and calendars).

Although Bluetooth and IrDA would seem to have considerable overlap, market adoption has shown that they can exist as complementary applications. IrDA has already found a niche for short-range, secure data exchanges when a broadcast signal is inappropriate. The release of products with both IrDA and radio-based networking capabilities will continue to be fueled by the IrDA-Data's capability to easily coexist with radio and its extension into application areas such as digital imaging and mobile communications.

Ultrawideband (UWB) Communications

The wireless networking systems we've discussed so far all use a small, well-defined set of frequencies that are regulated for power levels, emissions, and other characteristics that enable devices within those frequencies to peacefully coexist. Another emerging technology attempts to operate under the radar (so to speak) of these regulatory agencies. *Ultrawideband* (UWB) communication is a way of using a very wide frequency band at a very low power level in the regulated bands. UWB systems are derived from *Defense Advanced Research Projects Agency* (DARPA) research into military radar systems, but advances in UWB systems have enabled them to find possible applications in high-speed, short-range data communications, underground radar imaging, and secure communications. DARPA defines a UWB system as using a bandwidth greater than 25 percent of its center frequency, but literature also refers to systems using a bandwidth greater than 1.5 GHz as UWB. Power levels less than 200 mW restrict UWB systems to applications at less than 10 meters. UWB may be able to achieve speeds of 100 Mbps, with the aggregate speed of multiple systems in the same area much higher. So far adoption of this technology has suffered from regulatory hurdles and the lack of standardization. FCC Part 15 rules generally prohibit devices that emit an unacceptable level of signal. They allow what is called *incidental emissions*. The use of these very low power levels in UWB communications keeps emission levels low enough to qualify as incidental radio emissions under FCC Part 15 rules. The catch is that they are intentional (not incidental) emissions, and intentional emissions require FCC approval regardless of the power level. In May 2000, the FCC issued a proposal (FCC 00-163) to allow the unlicensed operation of UWB devices. The FCC review process of soliciting comments from interested parties before ruling is progressing. Because of the uniqueness of the technology and the application across a variety of licensed frequencies, the FCC process of public comment and review is still ongoing. In February 2002, the FCC adopted rules allowing limited marketing and operation of UWB devices. These uses include ground-penetrating radar, wall imaging, medical imaging, surveillance, and communication systems communication devices must operate in the 3.1-10.6 GHz range and be engineered for indoor use only. Several entities are investigating products for UWB, but widespread commercialization will be limited until the regulatory environment is clarified.

Wireless Local Area Networks (WLANs)

Home Radio Frequency (HomeRF) and the Shared Wireless Access Protocol (SWAP)

Targeting the home networking market, the *Home Radio Frequency* (HomeRF) Working Group was formed in 1998 to develop and promote specifications for the wireless networking of consumer products. The HomeRF Working Group is an industry organization whose original members included Compaq, HP, IBM, Intel, and Microsoft. By late 2000, its membership grew to nearly 100 members, but as other networking technologies gained popularity, its membership shrunk to approximately 70 members by late 2001.

In January 1999, the HomeRF Working Group released version 1.0 of the *Shared Wireless Access Protocol* (SWAP) specification. SWAP 2.0 was released in May 2001. Although SWAP is the name of the specification, HomeRF is frequently used to refer to systems employing SWAP. The terms HomeRF 1.0 and SWAP 1.0 are equivalent. Designed to handle voice and data applications, SWAP combines voice capabilities similar to DECT with data capabilities similar to the IEEE 802.11 frequency-hopping standard. Like 802.11 and Bluetooth, SWAP uses the 2.4 GHz frequency band. The network topology of SWAP supports both peer-to-peer networks and networks with central control points, which are similar to IEEE 802.11 APs.

SWAP 1.0 supports data rates of up to 1.6 Mbps and up to 4 full-duplex voice connections. Data transmissions employ the same basic techniques as IEEE 802.11. It uses a data packet collision avoidance protocol, has a frequency-hopping rate of 50 hops/sec and a range of approximately 50 meters, and its transmit power is 100 mW. For voice transmissions, it employs a TDMA technology like DECT. Because SWAP uses the unlicensed 2.4 GHz band, it doesn't suffer from the same frequency regulatory restrictions as DECT. Primarily because HomeRF chipsets did not have to support DSSS, *first-generation* (1G) HomeRF systems cost less than their IEEE 802.11 counterparts.

In 1999 and 2000, consumer products for the HomeRF 1.0 market were promoted by the major wireless networking vendors Intel and Proxim. The market for HomeRF equipment grew quickly, approaching almost $250,000,000 by the end of 1999, making HomeRF the first dominant wireless networking technology for the home. This began to change in 2000 as 11 Mbps IEEE 802.11b components began to fall in price. In addition, the FCC regulations for 2.4 GHz frequency-hopping systems specified that these systems had to use a minimum of 75 1-MHz-wide channels. The 1 MHz limitation made it impossible to offer higher speeds. With pressure from IEEE 802.11b in the marketplace, the HomeRF Working Group petitioned the FCC to modify the 2.4 GHz rules to allow wider channels and therefore faster speeds. In August 2000, the FCC changed the regulations to allow a minimum of 15 5-MHz-wide frequencies. This would enable SWAP 2.0 to support speeds comparable to IEEE 802.11b.

Market uncertainty about the regulatory environment and the delays in releasing the SWAP 2.0 specification started to erode HomeRF's market dominance. With dropping prices and speeds that were 5 to 10 times that of HomeRF, IEEE 802.11b equipment began to expand its market to include home networking systems as well as corporate applications. This market trend was validated in early 2001 as original HomeRF backer Intel said that it would base its future high-speed wireless networking products for both corporate and home on IEEE 802.11 standards. Although Intel continues to sell 1.6 Mbps HomeRF products, it has clearly thrown its support behind IEEE 802.11 standards, dealing a major blow to HomeRF in the marketplace. HomeRF's future depends on the availability of products and market acceptance for the 10 Mbps SWAP 2.0 that was released in mid-2001.

Even with the softening of Intel's support for HomeRF, some major backers are still promoting SWAP 2.0. These include Compaq, Motorola, National Semiconductor, Proxim, and Siemens. In addition, SWAP 2.0 has several new features besides higher data rates. It can operate at lower rates for an extended range and is backward compatible with the base of installed HomeRF equipment. It also supports eight full-duplex phone lines (up from four) and has added additional security, multimedia QoS capabilities, and roaming between APs. Proxim introduced the first 10 Mbps SWAP 2.0 products in October 2001. The HomeRF Working Group is optimistic about its future and has announced that HomeRF 3.0 will support data rates of 40 Mbps in 2002 and 100 Mbps in the future. A comparison of HomeRF 1.0 and HomeRF 2.0 is shown in Table 3-1.

High-Performance Radio Local Area Network (HIPERLAN) and HiSWAN

ETSI has been working on standards for radio-based wireless networks since 1991. These systems were designated *high-performance radio local area networks* (HIPERLANs). Like IEEE 802.11, the

Table 3-1

HomeRF 1.0 and HomeRF 2.0

	HomeRF 1.0	HomeRF 2.0
Frequency	2.4 GHz	2.4 GHz
Spread spectrum	Frequency hopping	Frequency hopping
Channel size	1 MHz	1 and 5 MHz
Line speed	1.6 Mbps	10 Mbps
Security	56-bit encryption	128-bit encryption

HIPERLAN standards define PHY and MAC layers. In 1996, ETSI published the HIPERLAN Type 1 standard, commonly referred to as HIPERLAN/1. HIPERLAN/1 operates in the 5 GHz frequency band and is designed to provide speeds up to 20 Mbps at a range of 30 to 50 meters. At the time of its release, the use of 5 GHz radio technology resulted in higher component costs and a shorter range than emerging 2.4 GHz systems like HomeRF and IEEE 802.11. As a result, HIPERLAN/1 was never accepted in the marketplace. ETSI recognized that HIPERLAN/1 was not a viable solution and in 1997 created a project called *Broadband Radio Access Network* (BRAN), which is commonly referred to as ETSI BRAN. ETSI BRAN is devoted to defining standards for broadband wireless systems that deliver data transmission speeds of greater than 25 Mbps. It has standards for several broadband wireless applications. HIPERMAN is a set of standards for broadband fixed wireless systems operating in the 2 to 11 GHz range. HIPERACCESS is a set of standards for high-frequency (40.5 to 43.5 GHz) broadband fixed wireless. HIPERLAN/2 is a WLAN standard that is designed to provide high-speed wireless access to *Asynchronous Transfer Mode* (ATM) networks, *third-generation* (3G) cellular networks, IEEE 1394, and IP networks. ETSI BRAN released the HIPERLAN/2 in February 2000. In Japan, the *Mobile Multimedia Access Communication* (MMAC) Project is developing the Japanese 5 GHz standard called *HiSWAN*.

HIPERLAN/2 has a PHY layer that is almost identical to IEEE 802.11a. It provides data transmission speeds up to 54 Mbps using *orthogonal frequency division multiplexing* (OFDM). Although the PHY layer is similar, frequency regulations are not consistent across the globe. European standards specify the 5.25 to 5.35 GHz band for indoor use at 200 mW power and the 5.47 to 5.725 GHz band at 1 W for indoor/outdoor use. The Japanese operate in the 5.15 to 5.25 GHz frequency band. In 1997, the FCC allocated 300 MHz of 5 GHz spectrum in three bands for high-speed data services. The FCC designated these bands as the *Unlicensed National Information Infrastructure* (U-NII). The three bands are 5.15 to 5.25 GHz, 5.25 to 5.35 GHz, and 5.725 to 5.825 GHz. As a move toward global frequency harmonization, the FCC is considering proposals to allow the 5.47 to 5.725 GHz band to be used by IEEE 802.11a in North America.

Another issue related to frequency harmonization is that European regulations are subject to local authorities as well. The local acceptance of the European regulations has been stalled by a possible conflict with radar that operates in the same band. The United Kingdom's Radiocommunications Agency is studying this issue and its results will affect the adoption of this technology by other countries. Resolution may not occur until late 2002. A major factor in the future growth of 5 GHz WLANs will be how and when global frequency regulations are stabilized. IEEE 802.11a established a market share advantage with the release of IEEE 802.11a equipment by Intel and others in late 2001. Until the European spectrum regulations for 5 GHz WLANs are finalized, equipment manufacturers will be hesitant to bring HIPERLAN/2 equipment to market. The HIPERLAN2 Global Forum, an industry group formed in 1999 by Nokia, Tenovis, Dell, Ericsson, Telia, and Texas Instruments, promotes HIPERLAN/2. Prototype systems were demonstrated in late 2000, but unlike IEEE 802.11a, HIPERLAN/2 components and equipment have yet to become commercially available.

Besides frequency regulations, the other main differences between IEEE 802.11a and HIPERLAN/2 lie in the MAC layer. The IEEE 802.11a MAC is designed to provide wireless IP. Other services such as QoS (IEEE 802.11e), security (IEEE 802.11i), and roaming (IEEE 802.11f) are specified separately. HIPERLAN/2 was designed with these capabilities. One fundamental difference is the use of fixed-length data packets (54 bytes) in HIPERLAN/2. IEEE 802.11a uses variable-length packets like other IP-related protocols. Another key difference is HIPERLAN/2's use of TDMA for medium access as opposed to the collision avoidance techniques of IEEE 802.11 specifications. HIPERLAN/2 has two radio management capabilities that differentiate it from IEEE 802.11a. It has a capability for controlling radio output called *transmission power control* (TPC). It also allows *dynamic frequency selection* (DFS). These capabilities help it avoid interference with other systems such as the U.K. radar mentioned previously. The IEEE 802.11h Task Group is working on a proposal to add DPC and DFS to IEEE 802.11a. The recommendations are in letter ballot approval stage and final approval is expected in early 2002. Like IEEE 802.11, HIPERLAN/2 has a network topology of APs acting as central control points for associated devices.

As a corollary to the IEEE 802.11 MAC, HIPERLAN/2 specifies a *Data Link Control* (DLC) layer that provides medium access and connection services. It is divided into several sublayers. These are the MAC, *Error Control* (EC), and *Radio Link Control* (RLC) layers. The RLC contains methods for *Radio Resource Control* (RRC), an *Association Control Function* (ACF), and *Datalink Connection Control* (DCC). The HIPERLAN/2 protocol stack is shown in Figure 3-2.

The HIPERLAN/2 MAC acts as a traffic cop for its associated devices. The MAC frame transmitted over the radio link is divided into time slots dedicated to specific functions such as uplink data, downlink data, and access control. The MAC layer analyzes requests for medium access and tells its devices when they should transmit in the MAC frame. The HIPERLAN/2 EC protocol handles transmission errors, retransmission requests, and the sequencing of data. The RLC provides the means to associate a device with an AP and establish the radio channels required for a connection, and allows control of the radio itself. The RLC ACF enables a device to request connectivity to an AP. This includes the exchange of network identification

Figure 3-2
HIPERLAN/2
protocol stack

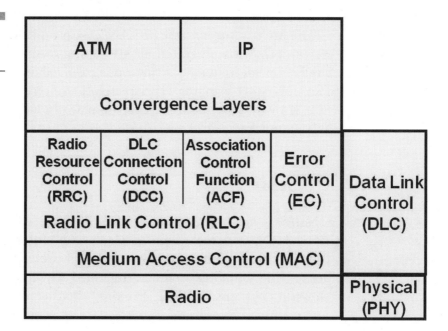

information, network capability information, and security information. This also includes the encryption algorithm required and the exchange of authentication keys.

The ACF is also used for disassociation procedures. Once a device is associated with an AP, the DCC function is used to set up a connection for the exchange of messages. Once a device is associated and a connection is established, the RRC protocol is used to control and exchange information about the radio link. This includes the exchange of signal-strength measurements that enable capabilities such as the handover of a mobile device between APs. This is also where DFS is handled. DFS enables an AP to change frequency due to environmental conditions or the presence of interfering signals. The RRC protocol also supports a power control function designed to let devices enter a low-power mode when inactive.

HIPERLAN/2's convergence layer enables connectivity to both packet-based protocols such as IP and cell-based protocols such as ATM. The convergence layer converts upper-layer service requests into service requests understood by the RLC layer. It also converts the data coming from the upper-layer to the fixed-size data format understood by the RLC layer. Just like any language translator, it converts both the words and the grammar. The first convergence layer defined is for IP packet handling.

The architecture of HIPERLAN/2 with built-in support for security and QoS is well suited for streaming media applications. Until the European frequency situation is resolved and equipment manufacturers start bringing HIPERLAN/2 equipment to market, IEEE 802.11a will continue to expand its place in the burgeoning 5 GHz WLAN market.

OpenAir

OpenAir is a 2.4 GHz frequency-hopping protocol based on Proxim's RangeLAN2 protocol. The Wireless LAN Interoperability Forum that was formed in 1996 promoted this protocol, but is no longer active. The founding members included ALPS Electric, AMP, Data General, Norand, NTT-IT, Proxim, Raytheon Electronics, Seiko,

Epson, and Zenith Data Systems. Specialized OpenAir products are still available through vendors such as Intermec and Proxim. With IEEE 802.11 supporting a frequency-hopping protocol, OpenAir systems are not as widely deployed today as general-purpose WLANs.

Wireless Metropolitan Area Networks (WMANs)

The *wireless metropolitan area network* (WMAN) encompasses a range of radio- and laser-based technologies targeted at providing wireless networking over distances of a few hundred meters to several miles. Wireless broadband, *broadband wireless access* (BWA), *wireless local loop* (WLL), fixed wireless, and wireless cable all refer to technologies for delivering telecommunications services over the last few miles of the network. Wireless broadband and BWA are general terms referring to high-speed wireless networking systems. WLL is derived from the wired telephony term *local loop*, which refers to the connection between a local telephone switch and a subscriber. WLL and fixed wireless generally refer to the delivery of voice and data services between fixed locations over a high-speed wireless medium, although there has been some talk of mobile applications of this technology. Fixed wireless includes *Local Multipoint Distribution Service* (LMDS), *Multichannel Multipoint Distribution Service* (MMDS), U-NII systems, and similar networks like AT&T's 1.9 GHz offering. Wireless cable usually refers to MMDS systems used to deliver television signals such as the *Instructional Television Fixed Service* (ITFS).

There are two basic network topologies supported by these systems. The simplest is a point-to-point system providing a high-speed wireless connection between two fixed locations. Bandwidth is not shared, but links typically require line of sight between the two antennae. The second topology is a point-to-multipoint network where a signal is broadcast over an area (called a *cell*) and communicates with fixed subscriber antennae in the cell. Because bandwidth in the cell is finite and is shared among all users, performance may be a concern in high-density cells. Systems of different frequencies may be combined to cover an area where terrain or other obstructions prevent full coverage.

Other than frequency, the main difference between fixed wireless systems and cellular, WLAN, and WPAN networks is the mobility of subscriber equipment. There has been some discussion about adding support for mobile subscriber equipment to fixed wireless systems. The addition of mobility support would enable these BWA systems to potentially function as *fourth-generation* (4G) cellular networks, delivering subscriber speeds of several megabits. There are still several technical, regulatory, and commercial hurdles to overcome before this could become a reality, but companies such as Wi-LAN have already started examining products targeted at this potential application.

With a worldwide subscriber base below 200,000, local wireless access solutions lag far behind other last-mile solutions like *Digital Subscriber Line* (DSL) and cable whose combined subscriber base is more than 10,000,000. Deployments outside of the United States have been more successful; both Asia and Europe have larger subscriber bases and growth rates. Several high-profile North American network operators have gone out of business or curtailed their operations in 2001 and 2002, making the near-term prospects cloudy. Future growth in the U.S. market will depend on more clarity in the frequency allocations for 3G services, the availability of user-friendly equipment, and general economic strength to stimulate demand.

Local Multipoint Distribution Service (LMDS)

LMDS is a fixed wireless, radio-based technology. *Fixed wireless* means that the transmitters and receivers don't move. In North America, LMDS operates in the 28 to 31 GHz frequency range, but may operate anywhere from 2 to 40 GHz in other regions. In 1998, the FCC held an auction for this spectrum, dividing each geographical area into an *A Block* and *B Block*. The A Block had a bandwidth of 1.5 GHz and the B Block had a bandwidth of 150 MHz. The intent was for the auction winners to deploy high-speed voice and data communications services in the last mile. The realities of deployment have not yet lived up to that vision.

The network topology of LMDS uses a central transmitter sending its signal over a cell with a radius of 5 km or less. Antennae are usually placed on rooftops for line of sight to the central transmitter. This is because 1G LMDS equipment uses radio technology that is affected by hills, walls, trees, and other physical barriers. This limitation may be lessened as equipment starts to adopt more advanced spectrum utilization techniques such as OFDM.

As a high-frequency outdoor radio technology, LMDS performance and range will vary depending on weather conditions. It has a range of less than 5 km and supports gigabit speeds, although services are usually offered at a much lower rate. The physics of the 30 GHz signal make it about a millimeter in length; this spectrum is sometimes referred to as *millimeter wave spectrum*. One effect of having such a small wavelength is that rain can effectively block the signal. In areas where rain is a factor, a lower frequency is required. A higher frequency allows faster data rates, but it also limits range, requiring more equipment to cover the same area as a lower-frequency technology. LMDS bandwidth in a specific area is shared among all the users like cable. To ensure end-user performance, networks must be built with excess capacity to handle sporadic peak loads and unexpected growth in the subscriber base. In addition, there are no standards governing LMDS implementations, leading to a number of incompatible proprietary solutions. Higher network deployment costs make 1G LMDS networks more suitable for high-margin business applications rather than residential use.

Several large telecommunication carriers began major North American network buildouts in the late 1990s. These included Winstar, Teligent, and XO Communications. Large capital outlays were required for the network buildout. These carriers had to assume large amounts of debt to fund their growth. This debt service burden coupled with slow subscriber growth and the deteriorating business climate of late 2000 and 2001 soon led to large operating losses. These financial difficulties have led Winstar and Teligent to file for Chapter 11 bankruptcy in early 2001, and XO has seen its stock price decline over 90 percent. Although these companies are still operating, debt restructuring and operational cutbacks have raised serious doubts about their ability to survive.

Multichannel Multipoint Distribution Service (MMDS)

MMDS is a fixed wireless, radio-based technology operating in frequency bands in the 2.1 to 2.7 GHz range. It is also referred to as the *Microwave Multipoint Distribution System*. Since the 1980s, MMDS has been used for a wireless cable TV broadcast service called ITFS. ITFS is primarily used by educational institutions and for rural cable access. Growing requests for licenses in the MMDS band caused the FCC to hold a spectrum auction in 1995. The Telecommunications Act of 1996 deregulated local access services, providing an opportunity for MMDS to become a larger part of the local telecommunications landscape.

One impediment to the greater use of MMDS spectrum for telecommunication services was that as a television broadcast medium, the spectrum's use was restricted to one-way transmission. In August 2000, the FCC allowed two-way transmissions. Operators were granted 72 MHz of spectrum across a 45 km cell. This provided an opportunity to bypass wired cable and incumbent telephone networks with a wireless system that could offer telephone, television, and high-speed Internet access. The potential for wireless data services at over 35 Mbps attracted the interests of Sprint and MCI WorldCom who invested over $2 billion to acquire MMDS licenses from existing carriers. Both began nationwide network rollouts.

Like LMDS, MMDS is deployed in a cellular topology using a fixed transmitter with an omnidirectional antenna. If not obstructed, the use of a lower frequency allows a range of over 50 miles. Obstructions such as hills and buildings can cause dead spots in coverage. MMDS is not susceptible to rain fade as the higher-frequency LMDS, but it is susceptible to a phenomenon called *multipath*, where radio waves bounce off objects—sort of a radio echo. Emerging technologies such as OFDM help combat this problem.

One drawback to MMDS is that an antenna must be installed at the subscriber's location. Current subscriber equipment generally requires installation by the carrier. This makes mass deployments costly, affecting subscription rates and carrier costs. Another issue that has affected MMDS is that 3G cellular service in Europe uses similar frequencies. Because of the lack of spectrum for 3G services in North America, there has been considerable pressure to reallocate these bands for 3G use in North America. The FCC ruled in late 2001 that MMDS license holders would not be obliged to surrender their licenses for 3G use, but allowed a portion of the frequency (2.5 to 2.69 GHz) to be used for 3G services. This left the door open to future regulatory action as the FCC indicated that there would be future rule-making regarding use of the new mobile frequency.

Despite the potential of MMDS, it has not yet been a commercial success. High deployment costs at the subscriber location and the large capital outlays necessary for network expansion have taken their toll in the economic downturn of 2001. Sprint's MMDS was targeted at residential users, and profitability concerns have caused Sprint to stop new installations. MCI WorldCom still offers its MMDS service, which targets small- and medium-sized businesses. AT&T is also curtailing its 1.9 GHz fixed wireless business. MMDS is still a viable technology and continues to grow internationally, but regulatory uncertainty and expensive subscriber equipment have hurt its growth in North America.

IEEE 802.16

In 1999, the 802 Committee set up the 802.16 Working Group to standardize approaches to BWA for metropolitan and other small regional areas. The IEEE 802.16 Working Group set up several Task Groups. The IEEE 802.16.1 Task Group was set up to establish a standard for point-to-multipoint systems operating in the 10 to 66 GHz frequency range. In February 2001, the Task Group designation was dropped and the standard began to be referred to as just IEEE 802.16. IEEE 802.16 systems support multiple user terminals communicating with a central base station. The base station serves as the connection to other networks, wired or wireless. It also supports the simultaneous operation of multiple services with different QoS requirements (for example, file transfers and video streams).

Although IEEE 802.16 focused on LMDS systems operating around 30 GHz, it is not restricted to that frequency. In December 2001, the IEEE 802.16 Wireless MAN Standard "Air Interface for Fixed Broadband Wireless Access Systems" was published.

The IEEE 802.16 PHY supports both a single-carrier air interface and a multicarrier OFDM interface. OFDM is also used in IEEE 802.11a. The OFDM Forum is working closely with the IEEE 802.16 Working Group to define the OFDM implementation in IEEE 802.16. A variation of OFDM called *vector orthogonal frequency division multiplexing* (VOFDM) has been promoted in the marketplace by the *Broadband Wireless Internet Forum* (BWIF). The BWIF is a trade organization that counts Cisco and Broadcom among its biggest advocates. The VOFDM systems promoted by the BWIF are a competing technology to IEEE 802.16 and differ mainly in their use of multiple antennae for signal transmission. It should be noted that after consideration of OFDM and VOFDM, ETSI BRAN has announced that it will adopt the OFDM PHY of the IEEE 802.16a standard and base its MAC on IEEE 802.16.

The IEEE 802.16 PHY also supports a frequency-based protocol called *frequency division duplex* (FDD) and a time division protocol called *time division duplex* (TDD). These are utilized to provide two modes of transmission based on application requirements. Modes for burst transmissions (such as file transfer) and continuous transmission (such as streaming video) are defined. Burst transmissions can use either FDD or TDD, whereas continuous transmissions must use FDD. Frequency division methods use separate frequency channels for uplink (terminal to base station) and downlink (base station to terminal) transmissions. Time division methods interleave uplink and downlink transmissions on the same frequency channel, using different time slots assigned by the base station. The PHY layer's support for a wide range of frequencies, channel widths, error correction schemes, and modulations results in a variety of possible distances and transmission speeds. This also provides the flexibility to operate in a variety of regulatory environments.

Three types of modulation are supported. *Quadrature phase-shift keying* (QPSK) supports speeds of up to 45 Mbps and operates over the longest distances. Two types of *quadrature amplitude modulation* (QAM) are supported: 16-QAM and 64-QAM. 16-QAM supports speeds approaching 90 Mbps, whereas 64-QAM supports speeds approaching 150 Mbps, but operates over shorter distances. Because it operates through the MAC layer, the IEEE 802.16 PHY layer is independent of the transport layers and therefore supports different transport mechanisms including ATM and IP.

The IEEE 802.16 MAC supports multiple PHY interfaces, and, like HIPERLAN/2, it provides flexibility in its transport protocol support through convergence layers for ATM and IP. IEEE 802.16 has also utilized many of the concepts specified in the *Data over Cable Service Interface Specification* (DOCSIS) standard for security and QoS. The IEEE 802.16 MAC provides security with support for authentication mechanisms, such as *Rivest, Shamir, and Adleman* (RSA) private keys and X.509 certificates, and *Data Encryption Standard* (DES) encryption. The baseline IEEE 802 MAC layer is based on the Ethernet protocol. In order to support ATM and IP as well as Ethernet, the IEEE 802.16 MAC was reworked to make it transport independent. Transport-specific functions are handled in the ATM and IP convergence layers. Both convergence layers support QoS and protocol-specific header suppression. The ATM layer supports *switched virtual circuit* (SVC) and *permanent virtual circuit* (PVC) operation. The IP convergence layer supports both *IP version 4* (IPv4) and *IP version 6* (IPv6).

The IEEE 802.16 MAC also provides a flexible connection mechanism between the base station and subscriber station. The base station grants connections to the subscriber station on either a connection or terminal basis. Granting bandwidth to a connection allocates bandwidth to a specific path and is suitable for single-user terminals. The management of bandwidth by the base station uses base station resources, but allows the use of simpler low-cost terminals. Granting bandwidth to a terminal provides the terminal with a pool of bandwidth that it may allocate to multiple connections. This enables the terminal to manage bandwidth and QoS for each application connection as needed. Offloading bandwidth management to the terminal uses fewer base station resources, but it requires more sophisticated and costly terminal equipment.

The IEEE 802.16 Working Group looked to extend the 10 to 66 GHz standard to the licensed and unlicensed bands in the 2 to 11 GHz frequency band. They initially formed two separate Task

Groups. Task Group *a* (IEEE 802.16a) is working on an amendment to IEEE 802.16 for the 2 to 11 GHz licensed bands. The unlicensed bands in the 2 to 11 GHz range were initially handled by Task Group *b* (IEEE 802.16b). In November 2001, IEEE 802.16b was absorbed into IEEE 802.11a. IEEE 802.11a is now chartered with providing an amendment for both licensed and unlicensed frequencies in the 2 to 11 GHz band. IEEE 802.16a is now a draft standard and is scheduled for completion in mid-2002.

In September 2001, the recommended practice IEEE 802.16.2, "Coexistence of Fixed Broadband Wireless Access Systems," was published. This document describes deployment parameters and operational procedures that enable adjacent or overlapping networks in the 10 to 66 GHz frequency band to manage interference. It recommends antenna patterns, power levels, separation distances, and receiver sensitivity to noise caused by other fixed wireless systems. The IEEE 802.16.2 Working Group is working to extend the recommended practice to the 2 to 11 GHz frequency band. It has formed the IEEE 802.16.2a Task Group for this purpose. The IEEE 802.16.2a Task Group is also investigating practices related to the coexistence of point-to-point 2 to 11 GHz networks with the point-to-multipoint topology. Table 3-2 lists the Task Groups of IEEE 802.16.

HIPERMAN and HIPERACCESS

ETSI BRAN has three major projects related to WMAN networking. HIPERACCESS covers the higher frequencies and is optimized for the 40.5 to 43.5 GHz band. Like IEEE 802.16, it is a point-to-multipoint system. It is designed to support speeds up to 25 Mbps and uses TDMA as the air interface. HIPERMAN covers the 2 to 11 GHz frequency range and allows both point-to-point and mesh configurations. HIPERLINK is designed to provide short-range (150 meters), wireless point-to-point connections at speeds up to 155 Mbps. It operates in the 17 GHz frequency band. These specifications are still under development, with the HIPERACCESS specification closest to completion. ETSI BRAN and the IEEE have been cooperating closely on coordinating their respective WMAN standards efforts. They have each appointed official liaisons that have

Table 3-2

802.16 task groups

802.16 (formerly 802.16.1)	Air Interface for Fixed Broadband Wireless Access Systems 10 to 66 GHz Published December 2001	
802.16a	Air Interface for Fixed Broadband Wireless Access Systems Amendment Licensed and unlicensed 2 to 11 GHz Expected completion mid-2002	
802.16b	Air Interface for Fixed Broadband Wireless Access Systems Amendment Unlicensed 2 to 11 GHz Absorbed into 802.16a in November 2001	
802.16.2	Coexistence of Fixed Broadband Wireless Access Systems 10 to 66 GHz Published September 2001	
802.16.2a	Coexistence of Fixed Broadband Wireless Access Systems 2 to 11 GHz including point to point In progress	

access to the standards groups in each organization. Evidence of this cooperation is the adoption of the IEEE 802.16 MAC and OFDM PHY by the HIPERMAN group.

Free-Space Optics (FSO)

Just as infrared provides an optical alternative to radio for WPAN and WLAN networks, *free-space optics* (FSO) provides an alternative to radio-based WMANs. FSO has been around for 20 years or so, but line-of-sight restrictions and its susceptibility to rain, fog, and other environmental conditions have limited its adoption. It does have some advantages such as speedy low-cost installation, operation in an unlicensed frequency, and the security of a nonbroadcast signal. Recent technical advances have allowed FSO to find wider acceptance, and it is increasingly being seen as a viable alternative where radio or optical fibers are inappropriate or unavailable.

FSO uses lasers to transmit data through the air. It can provide connectivity at very high data rates, with some vendors claiming transmission rates up to 10 gigabits. FSO systems can be deployed in point-to-point, point-to-multipoint, or mesh configurations. All of these configurations require line of sight between communicating nodes. Typical applications are high-speed building-to-building links, backhaul of traffic from access systems to network hubs, or disaster backup capabilities. FSO's ease of installation makes it well suited to ad hoc or temporary connections. Indeed, the ability to set it up quickly enabled it to be used for emergency high-speed connectivity after the attack on the World Trade Center on September 11, 2001.

The major drawback to FSO is the requirement for a clear line of sight. Performance and reliability can degrade quickly in adverse weather conditions. This generally restricts the effective distance to 1 km or less. Spurious blockages such as birds can also affect the transmission. Some vendors such as Optical Access use a microwave backup as part of their equipment to provide higher availability in adverse weather. Mesh topologies also mitigate blocked transmission problems, but may be more costly to install. FSO systems are generally installed on rooftops or in windows. Because tall buildings sway, maintaining alignment of the laser nodes can be a problem, but recent advances in autotracking has made this less of an issue. The major drawback to rooftop installations is economic. It requires negotiated access and maintenance contracts with building owners. These can be costly and complicated.

Weighing the advantages against the disadvantages, FSO has a solid future as the demand for broadband access grows. Industry studies project the market for FSO equipment and services to be over $1 billion by 2004.

Regional and Global Wireless Networks

Until the 1980s, regional and global wireless networks were limited to voice transmission. Original cellular telephony technologies were analog and not engineered for digital data. Networks were generally regional in nature and each country had different standards. Roaming and interoperability were problems. As the demand for mobile voice services grew, technical advances and standardization began to enable the *second-generation* (2G) of cellular phone service.

The primary technical advance for 2G networks was the introduction of digital transmission technologies. These included TDMA, *Code Division Multiple Access* (CDMA), and the *Global System for Mobile Communications* (GSM). These technologies set the stage for the introduction of the first wireless data services and the inevitable transition to higher speeds and ubiquitous coverage. The term TDMA can be confusing. It is sometimes used to refer to the general technique of using different time slots on a frequency band to multiplex uplink and downlink traffic as well as to refer to specific implementations of the technique. In North America, 2G TDMA systems are based on the IS-136 standard. Japan has its own version of 2G TDMA cellular called *Personal Digital Cellular* (PDC). In other regions, the dominant TDMA technology is GSM. CDMA, on the other hand, does not use time slots to interleave transmission blocks, but it uses a digital marker as a separator. CDMA is the underlying technology of IS-95A.

Responding to the hodgepodge of standards and systems in Europe, ETSI produced the GSM standard and mandated its use across Europe. Fueled by an increasingly mobile world and a widely implemented standard, GSM spread quickly and now has hundreds of millions of users across the globe. Although a few GSM systems were deployed in North America, North American 2G cellular is primarily IS-136 or IS-95A. Although there are IS-136 and IS-95A systems deployed outside of North America, GSM is by far the dominant global 2G cellular technology. The 1990s saw the mass commercialization of digital cellular technologies. Although they are digital in nature, 2G systems are primarily used for voice services with only limited data capabilities. As the Internet fueled demand for mobile data services, standards for cellular networks capable of both voice and data began to evolve. These networks are referred to as 3G cellular.

The path to future high-speed cellular networks was outlined by the *International Telecommunications Union* (ITU) in its *International Mobile Telecommunications 2000 Project* (IMT-2000). This project was started in 1992 to provide the framework for a global wireless communications network. The considerable delays in reaching a global consensus on a standard resulted in a compromise in 1999 with three different modes of operation (two CDMA and one TDMA). The first commercial deployments of 3G technology occurred in 2001, but widespread deployment has been gradual. The large installed base of 2G networks, the technical challenges of high-speed cellular data, and the length of time required to standardize and implement 3G networks opened a window for the introduction of transitional network standards. These transitional standards typically add data capabilities to existing cellular infrastructure, but data speeds are lower than 3G. These transitional technologies are referred to as 2.5G cellular, and several were deployed worldwide by the end of 2001. Figure 3-3 shows the typical migration paths from 2G to 3G. Intermediate steps may be skipped in some deployments.

2.5G Cellular Networks

There are several technologies that qualify as 2.5G. These include *General Packet Radio Service* (GPRS), IS-95B, and iMode. GPRS adds packet data capabilities to GSM networks. It uses the existing GSM infrastructure for voice and adds a set of protocols and network

Figure 3-3
Cellular technology
evolution

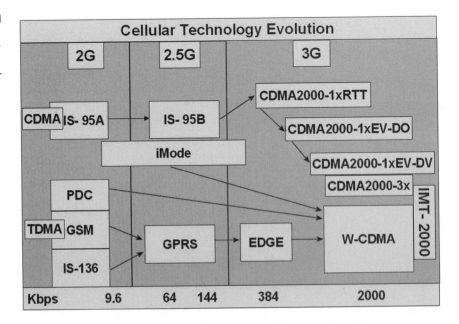

elements supporting packet data. GPRS supports data rates up to 171 Kbps, with deployed rates generally lower. A variation of GPRS called *Enhanced Data Rates for GSM Evolution* (EDGE) provides data rates up to 384 Kbps, but uses a different radio interface than GPRS. One drawback is that this requires operators to use a portion of their voice spectrum for EDGE data services. GPRS networks are already active in many countries across the globe. In late 2001, Cingular affiliate Salmon PCS and VoiceStream announced plans for future EDGE deployments. Table 3-3 lists some major GPRS network operators around the world. IS-95B added 64 Kbps data capabilities to IS-95A. It is deployed in Korea and Japan, but most North American IS-95 operators have skipped IS-95 and begun implementing CDMA-2000 3G networks.

3G Cellular Networks

In 2001, the first CDMA-2000 and W-CDMA 3G cellular networks were deployed. CDMA-2000 networks have a somewhat easier path to implementation because CDMA-2000 is designed to work on existing frequencies. Wideband CDMA (W-CDMA) systems were first developed by NTT DoCoMo in Japan. W-CDMA systems are designed as the successor to GSM/GPRS systems. They include several radio variations utilizing 5 GHz of bandwidth. It operates on the 450, 700, 800, 900, 1,700, 1,800, and 1,900 MHz bands. With a carrier frequency spacing of 1.25 MHz, it uses existing spectrum more efficiently than W-CDMA, which requires new spectrum and has a carrier spacing of 5 MHz.

The first CDMA-2000-1x systems were deployed in Korea in late 2000. Korean operators SK Telecom, LG Telecom, and KT Freetel now have more than 2.5 million subscribers. CDMA-200 is the name for the IMT-2000 multicarrier version of CDMA. CDMA-2000-1x provides voice and data rates of over 300 Kbps per carrier. The next version of CDMA-2000 is called CDMA-2000-1xEV-DO, or CDMA-2000 Evolution for Data Only. CDMA-2000-1xEV-DO uses a dedicated 1.25 MHz carrier for data and provides data rates of 2.4 Mbps. The next step offers the same data rates and integrates voice and data on the same 1.25 MHz carrier. Like W-CDMA, CDMA-2000-3X uses a 5 MHz carrier for integrated

Table 3-3

GPRS network
operators

Country	Carrier
Australia	Telstra Cable & Wireless Optus
Austria	Connect Austria Max Mobile Mobilekom T-Mobil
Canada	Microcell
Germany	D2 Vodaphone E-Plus Mobilefunk Viag Interkom
Hong Kong	Hutchison New World SmarTone PCCW
Ireland	EirCell
Italy	Blue Omnitel Wind
Lebanon	Cellis
Netherlands	Libertel KPN
Russia	KB Impuls
Singapore	SingTel Starhub
Spain	Airtel Amena
Sweden	Telia Europolitan
Switzerland	Swisscom Sunrise
Taiwan	KG Telecom
United Kingdom	BT Cellnet One2One Vodaphone
United States	AT&T Wireless Cingular VoiceStream

voice and data. In the United States, Sprint, Verizon Wireless, and Leap Wireless have all started trials or limited rollouts of CDMA-2000-1x networks.

Throughout the summer of 2001, Manx Telecom and NTT DoCoMo raced to deploy the world's first W-CDMA network. Both launched commercial networks in the fall of 2001. Manx Telecom launched its service on the Isle of Mann in the United Kingdom. NTT DoCoMo launched the *Freedom of Mobile Multimedia Access* (FOMA) service to a limited set of subscribers in the Tokyo area. Videophone handsets for the service sold out in the first week. NTT DoCoMo expects 150,000 subscribers in the first quarter of 2002. Japan's J-Phone is scheduled to launch W-CDMA service in 2002. Table 3-4 shows some of the first 3G networks worldwide.

Despite the efforts of the ITU to promote worldwide standards, cellular networks will continue to be a mixture of technologies for the foreseeable future. Although carriers have paid large licensing fees for 3G spectrum, consumer demand is still nebulous. In Europe

Table 3-4

3G networks

Country	Carrier	Technology
Japan	NTT DoCoMo	W-CDMA
Japan	J-Phone	W-CDMA
United Kingdom (Isle of Mann)	Manx Telecom	W-CDMA
Korea	SK Telecom	CDMA-2000-1x
Korea	LG Telecom	CDMA-2000-1x
Korea	KT Freetel	CDMA-2000-1x
United States	Verizon	CDMA-2000-1x
United States	Sprint PCS	CDMA-2000-1x
United States	Leap Wireless	CDMA-2000-1x
United States	U.S. Cellular	CDMA-2000-1x
Brazil	Telesp Celular	CDMA-2000-1x

alone, carriers paid over $100 billion for 3G licenses in 2000. Everyone seems to agree that wireless Internet access and other wireless data applications such as mobile commerce are enticing, but the cellular industry has yet to prove economic viability, given the huge investment required.

Everyone seems to be searching for the killer app that will drive consumer demand. The deployment of a variety of 2.5G and 3G technologies has put pressure on cellular handset manufacturers to support multiple air interfaces and application software. Cellular handsets, PDAs, and portable multimedia electronics are combining to enable new services. Personal wireless devices of the future will have a combination of WPAN, WLAN, and cellular network capabilities. Semiconductor manufacturers are developing chipsets combining multiple technologies in support of this trend. Consumer devices combining Bluetooth, IEEE 802.11, and cellular technologies are now coming to market.

After the explosive growth of cellular voice services in the 1980s and 1990s, the cellular industry is struggling to find its footing in the data-centric new millennium. This has provided an opportunity for other technologies such as IEEE 802.11 to establish itself as a proven wireless access method at the local level. This has led to the emergence of IEEE 802.11 wireless carriers who can provide wireless Internet access and other data services at speeds of 11 Mbps today (soon to be 54 Mbps). Some in the cellular industry have seen this as a threat to the deployment of 3G technologies. It is more likely that alternative wireless access methods will complement cellular networks. This has already started with the promotion of IEEE 802.11 network access as a service offered by GSM network operators. WLAN hotspots in hotels, airports, and other public areas are being created and maintained by GSM carriers. Access is billed through today's GSM infrastructure, and current roaming agreements provide access outside of a carrier's home area. This is discussed in detail in the section on IEEE 802.11 carriers in Chapter 5.

Short Messaging Service (SMS), iMode, and Cellular Digital Packet Data (CDPD)

NTT DoCoMo started a CDMA-based wireless Internet service called iMode in February 1999. A little over a year later the service had over 6 million subscribers. Designed to deliver e-mail, online banking, and information services like news, weather, and traffic information to a cellular phone, the service has been an unqualified success in Japan. Operating at 9.6 Kbps, content is kept simple and users pay for data packets, not for connection time. NTT DoCoMo partners with content providers and helps them build iMode web sites that can then charge for access.

Short Messaging Service (SMS) is a wireless data service for sending and receiving short text messages (usually less than 200 characters) on a cellular phone. It has been the most successful wireless data service with global cellular networks processing up to 20 billion messages per month. About 80 percent of this traffic is generated in Europe and Asia. SMS is a slow-cost, simple text messaging facility that lets users address the cell phone through a standard e-mail address. Although addressing is through an e-mail address, SMS is not the same as e-mail. Messages are sent through the cellular network's control frequencies and don't use valuable voice spectrum. A store-and-forward message switch provides delivery. This enables messages to be stored and delivered when the phone is available. SMS is available on most 2G and 2.5G cellular networks.

Given the success of SMS and looking to take advantage of the higher data rates offered by 2.5G and 3G networks, Nokia, Ericsson, and other leading cellular network vendors are now promoting a new version of SMS called *Multimedia Messaging Service* (MMS). MMS increases the amount of text that can be sent and adds multimedia capabilities such as the capability to send graphics and audio files. One of the key success factors of SMS is its low cost. MMS is by definition a heavier consumer of bandwidth and as such it is more costly for the carrier. It will also require new handsets that are capable of MMS applications. The economics of delivering rich data content on a large scale over 2.5G and 3G networks is yet to be proven, but the carriers have a large existing set of customers who are prime candidates for an upgrade.

Cellular Digital Packet Data (CDPD) is a 19.2 Kbps wireless data service. CDPD uses idle time between voice transmissions to send and receive data on 2G networks. Similar to connecting over a standard phone line, it requires a CDPD modem that allows frequency sharing. From a carrier's perspective, it requires specialized network equipment. CDPD has been around since the early 1990s, but is being overwhelmed by newer, faster digital technologies.

As the worlds of data and wireless communications merge, wireless data has become one of today's fastest growing technologies. The lines between WPANs, WLANs, WMANs, and 3G cellular are quickly blurring from a user's perspective. Hardware vendors are releasing products supporting multiple technologies such as GPRS/802.11 and Bluetooth/802.11 cards. IEEE 802.11 has emerged as a technology that is not only useful in the local area, but as a bridge between other wireless data technologies.

Trade Groups
and
Chipmakers

Industry and Educational Organizations

With every emerging technology, it is important to have an understanding of the organizations participating in the marketplace. Industry, educational, and professional organizations provide industry-wide services, drive technical innovation, and accelerate commercial adoption. This section discusses the key organizations in the 802.11 community.

The Wireless Ethernet Compatibility Alliance (WECA) and Wi-Fi™

The most influential 802.11 industry organization is the *Wireless Ethernet Compatibility Alliance* (WECA). It was formed in 1999 by Intersil, Cisco (acquired Aironet), Agere Systems (spun off from Lucent), 3Com, Nokia, and Symbol Technologies. WECA has been very successful, and with the popularity of 802.11b, its membership is approaching 150. Microsoft became a board member in July 2001, and Intel joined the board as well in September 2001.

WECA was established to promote 802.11b through the development of an interoperability certification procedure. As we have seen, the 802.11 series of standards provides a significant level of flexibility in implementation. The resulting lack of interoperability among the first implementations of 802.11 products limited consumer choice and was an impediment to market acceptance. The founding members of WECA wanted to remove this impediment to the adoption of 11 Mbps 802.11b networks.

To accomplish this goal, WECA began certifying equipment for interoperability in April 2000. Equipment certified by WECA was given the Wi-Fi seal of approval. Wi-Fi stands for *Wireless Fidelity*. Because of its commercial success, the Wi-Fi designation has become better known than WECA. The Wi-Fi seal of interoperability ensures that *access points* (APs), *network interface cards* (NICs), and other 802.11b equipment will function together—even if it is manufactured by different vendors. The success of this program is measured by the fact that since its inception, more than 75 vendors have certified over 300 products. In November 2001, WECA announced that it would be performing the certification of 5 GHz IEEE 802.11a products under the name Wi-Fi5. The *Agilent Technologies' Interoperability Certification Lab* (AICL) provides the testing services for the IEEE 802.11b Wi-Fi certification program.

The AICL was launched in 1998 as a private test lab for communications products and was called the *Silicon Valley Network Lab* (SVNL). It offered testing services for a number of communications products including Ethernet switches, routers, *Digital Subscriber Line* (DSL) equipment, and 802.11b equipment. Agilent Technologies acquired SVNL in late 1999 and continued to provide WECA with Wi-Fi testing services. As a major manufacturer, Agilent's acquisition initially raised concerns about possible conflicts of interest, but the AICL has focused on working with industry groups such as WECA rather than individual equipment vendors. AICL does not market itself for comparative vendor testing and has continued to successfully certify equipment from a variety of vendors. As 802.11 adoption has grown, certification of equipment has expanded from NICs, APs, and gateways to integrated network interfaces and networked peripheral products such as print servers. Table 4-1 lists companies that have Wi-Fi-certified equipment.

Table 4-1

Wi-Fi-certified equipment

Company	Wi-Fi Products
2Wire	Gateways
3Com	Network Cards, Access Points, Gateways
Accton	Network Cards, Gateways
Acrowave Systems	Network Cards, Access Points
Actiontec	Network Cards, Gateways
Agere	Network Cards, Access Points, Gateways
Ambit	Network Cards, Access Points
Apple	Network Cards, Access Points
ARESCOM	Access Points, Gateways

Table 4-1 cont.

Wi-Fi-certified
equipment

Company	Wi-Fi Products
Askey	Network Cards, Access Points
Atmel	Network Cards
Avaya	Network Cards, Access Points
BroMax	Network Cards, Access Points
Cisco	Network Cards, Access Points
Colubris Networks	Network Cards
Compaq	Network Cards, Access Points
CREWAVE	Network Cards, Access Points
D-Link	Network Cards, Access Points
Dell	Network Cards, Access Points
Delta Networks	Network Cards
ELSA	Network Cards, Access Points
EMTAC	Network Cards, Access Points
Enterasys	Network Cards, Access Points
Epson America	Print Servers
Fine Digital	Access Points
Fujitsu	Network Cards, Access Points
Galtronics	Access Points
Gateway	Integrated Network Cards
GemTek	Network Cards, Access Points
Global Sun	Network Cards, Access Points
Hewlett Packard	Network Cards, Print Servers
Hitron Technologies	Print Servers
IBM	Network Cards, Access Points, Gateways
Intel	Network Cards, Access Points, Gateways
Intermec	Network Cards, Access Points
Intersil	Network Cards
Inventec	Network Cards, Access Points
IP One	Network Cards, Access Points
Kyocera	Network Cards, Access Points

Table 4-1 cont.

Wi-Fi-certified equipment

Company	Wi-Fi Products
LG Electronics	Network Cards, Access Points
Linksys	Network Cards, Access Points
Melco (Buffalo)	Network Cards, Access Points
MITAC	Network Cards
Mitsumi	Access Points
MMC Technology	Network Cards, Access Points
NEC	Network Cards, Access Points
NETGEAR	Network Cards, Access Points
NextComm	Network Cards
Nokia	Network Cards, Access Points
Nova Technology	Network Cards, Access Points
NTT-ME	Network Cards
OTC Wireless	Network Cards
Philips	Network Cards
Proxim	Network Cards
Psion Teklogix	Gateways
Quanta Computer	Network Cards
RFTNC	Network Cards
Samsung	Network Cards, Access Points
Sharp Electronics	Network Cards, Access Points
Siemens	Network Cards, Access Points
SMC Networks	Access Points
Sony	Network Cards
Symbol Technologies	Network Cards, Access Points
TOKO	Network Cards, Access Points
Toshiba	Network Cards, Gateways
Troy Wireless	Network Cards, Print Servers
Universal Scientific Industrial	Network Cards
Wistron	Network Cards, Access Points
Yamaha	Access Points
Z-Com	Network Cards

The University of New Hampshire InterOperability Lab (IOL)

Another organization involved in interoperability testing for 802.11 equipment is the University of New Hampshire's *InterOperability Lab* (IOL). The IOL has a number of different consortia involved in developing conformance test suites and providing a forum for vendor interoperability testing. These consortia cover a wide range of communications technologies, including 802.11. The Wireless (802.11) Consortium members are Agere, Anixter, Atmel, Cisco, Conexant, Enterasys, Intermec, Intersil, Mobilan, netIQ, Network Associates, Sharewave, Wireless Solutions, and Zoom. The members pay an annual membership fee and contribute equipment to the IOL. In return, they have a neutral forum for testing with other consortium members, help develop conformance test suites, and gain valuable exposure for their products in the academic and research community. Unlike AICL, the IOL does not certify equipment; rather, it provides a way to validate early designs and equipment incorporating emerging technologies.

The Wireless LAN Association (WLANA)

The oldest 802.11 industry group is the *Wireless LAN Association* (WLANA). The WLANA was formed in 1996 by the IBM Corporation, Cisco (acquired Aironet), Agere Systems (spun off from Lucent), 3Com Corporation, Advanced Micro Devices, Compaq (merged with DEC), Intersil (spun off from Harris Semiconductor), Intermec (acquired Norand), Proxim, Inc., Symbol Technologies Inc., Raytheon Electronics, and Windata. These 12 companies formed the WLANA as a nonprofit educational group dedicated to promoting the adoption of 802.11 and now includes over 30 members. Originally called the Wireless WLAN Alliance, it changed its name to the Wireless LAN Association in May 2000. The WLANA provides a clearinghouse for information about 802.11 and related networking technologies.

The Orthogonal Frequency Division Multiplexing (OFDM) Forum

The OFDM Forum was formed to promote the use of *orthogonal frequency division multiplexing* (OFDM) in a variety of wireless applications and as such their work includes 802.11. Their work is divided into three main areas of interest—WLANs (802.11), fixed wireless (*Multichannel Multipoint Distribution Service* [MMDS], *Local Multipoint Distribution Service* [LMDS], and 802.16), and mobile broadband (*third-generation/fourth-generation* [3G/4G] cellular). OFDM is the transmission mechanism for the *physical* (PHY) layer in 802.11a and 802.11g. It is also a key technology proposed for adoption in the 802.16 MAN standard. The OFDM Forum actively promotes a variety of OFDM called *wideband OFDM* (W-OFDM). Wi-LAN, Philips Semiconductors, Nokia, Ericsson, Samsung, Sony, and the Caltrans Testbed Center for Interoperability at the University of California at Santa Barbara founded the OFDM Forum in December 1999. It now has over 50 members. The OFDM Forum is an active participant in the standards-setting process, particularly with the IEEE 802 LAN/MAN Committee and the *European Telecommunications Standards Institute/Broadband Radio Area Network* (ETSI BRAN) *high-performance radio local area network* (HIPERLAN/2) group.

802.11 Chipmakers

This section discusses the companies that are involved in producing semiconductor products for the 802.11 market. These companies exhibit different levels of vertical product line integration (chips to boards to end-user products), but all are involved in producing the basic semiconductor components of 802.11 equipment with 2001 sales of over $200 million, the WLAN semiconductor market is estimated to grow at a compound annual growth rate approaching 30 percent over the next few years and could be a $1 billion market by 2005, with a major portion based on 802.11 technology. Led by Intersil and Agere (Lucent), the 802.11b semiconductor market is still growing, but will likely reach maturity over two to three years. The 5 GHz market, on the other hand, is poised for rapid growth. Despite the unfavorable economic climate of 2002, there has been an influx of investment, several acquisitions, and the emergence of a number of startups.

The following semiconductor products provide the PHY and/or *Medium Access Control* (MAC) functionality required by 802.11. Several major functions are required, and each of the following manufacturers has taken its own approach. At the most basic level, chipset solutions can be logically divided into analog functions providing the radio interface and digital functions providing the computer interface. The radio interface provides transceiver functionality and includes components like antenna interfaces, power amplifiers, low-noise amplifiers, and filters. Many times the analog signal is converted to an *intermediate frequency* (I/F) before being passed to the digital processor. When the radio function passes the signal to the digital processor without an I/F conversion, it is called a *direct down conversion*. The digital processing function may include a combination of baseband signal processing, a *Reduced Instruction Set* (RISC) *central processing unit* (CPU) for MAC processing, *input/output* (I/O), and bus interfaces like the *Personal Computer Memory Card International Association* (PCMCIA), *Peripheral Component Interconnect* (PCI), and/or *universal serial bus* (USB). It may also include specialized processors for security or *quality of service* (QoS). Basic chipset functions are shown in Figure 4-1.

Figure 4-1
Basic WLAN chipset functions

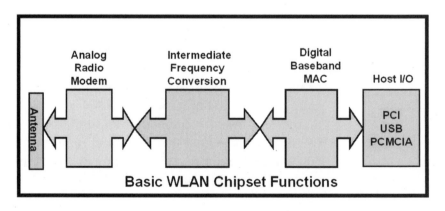

Basic WLAN Chipset Functions

Manufacturers are differentiated by their choice of semiconductor materials, firmware, design architecture, development tools, and performance parameters like power consumption, transmission speed, and distance. These designs generally consist of a variety of fixed-function components, but there is another approach that utilizes highly programmable processors.

Two of the most commonly used terms for programmable processor technology are *system-on-a-chip* (SoC) and *field programmable gate array* (FPGA). These technologies allow the programming of the processor logic as opposed to hardwiring it in the silicon. Although the trade-off for flexibility is usually cost and implementation complexity, these vendors generally offer programming tools and design expertise as well. SoC and FPGA are general terms and are not limited to 802.11 or wireless applications. Another related term is *Software-Defined Radio* (SDR). This is a PHY processor that is capable of being programmed to handle multiple radio technologies. SDRs may be implemented using FPGA or SoC technology. Recent improvements in the performance of FPGA *digital signal processors* (DSPs) have enabled their application to the processing of high-frequency radio signals.

Intersil

Claiming over 60 percent of the WLAN semiconductor market, Intersil is a leading manufacturer of *integrated circuit* (IC) solutions for the 802.11 market. Originally part of Harris Semiconductor, it became a separate entity in August 1999. In addition to products for the 802.11 market, its product lines include power management ICs and analog/mixed signal ICs. Its most successful product has been the PRISM® line of solutions for the 802.11b market. Intersil designs and manufactures all the major elements, including both radio- (PHY) and MAC-level components. Its *radio frequency* (RF) and power amplifier components utilize *silicon germanium* (SiGe), whereas its MAC-level components use *complementary metal oxide semiconductor* (CMOS) processes. It has produced three generations of 2.4 GHz chipsets, with each generation having fewer chips. Because it designs both the radio and baseband/MAC chips, Intersil has tightly integrated 802.11 solutions. A good example of this is its two-chip PRISM 3® solution. Intersil has combined the I/F and filtering functions on a single chip that operates in conjunction with a complementary baseband/MAC chip.

Although a dominant player in the 802.11b market, the emerging market for 5 GHz 802.11a and HIPERLAN/2 has provided an opportunity for newer players such as Atheros. Intersil has responded with an announcement of its own 5 GHz offering. In April 2001, Intersil acquired No Wires Needed. No Wires Needed produced an AP, Linux products, and a high-performance ARM MAC chip supporting 54 Mbps operation. In June 2001, Intersil announced its 5 GHz product line—PRISM Indigo. Targeted at the 802.11a market, the chipset includes four chips: an SiGe power amplifier, an SiGe radio/I/F chip, a BiCMOS I/F/baseband chip, and a CMOS baseband/MAC chip with an ARM9 RISC processor. The complete chipset is priced at $35 in quantities of 100,000. Intersil has also indicated that the chipset will support 802.11e QoS for multimedia applications. Intersil's 5 GHz chipsets were available for sampling in early 2002.

In September 2001, Intersil announced the availability of an AP reference design based on the PRISM 2® chipset. What is unique about this is that the MAC is implemented on ARM Holdings, plc ARM9 processor running Linux. Because the Linux open-source community provides a wide range of applications, it enables the faster development of APs with features such as enhanced security. In January 2002, Intersil announced a chipset for IEEE 802.11g. This chipset includes support for CCK and OFDM operation. Cisco and Intersil have announced a collaboration on a joint reference design for IEEE 802.11g. Intersil will supply PHY technology with Cisco supplying the MAC.

Intersil's networking products are used in a wide variety of products, and PRISM has been incorporated into over 100 WLAN products from over 50 vendors. Among its customers are Intel, Alcatel, 3Com, Compaq, Cisco, Dell, HP, IBM, Nokia, Sony, Sharp, and Toshiba. Intel has incorporated Intersil's 802.11b PRISM 2.5 chipset in the Intel PRO/Wireless 2011B LAN products. They also cooperated in developing a solution for Compaq's Evo line of notebooks. This solution uses Compaq's MultiPort module to provide both 802.11b and Bluetooth capability. The MultiPort module is a self-contained unit with all the wireless connectivity hardware, including the antenna, inside the module, eliminating a protruding antenna. IBM utilizes Intersil's PRISM 2.5 in its new ThinkPad® T23 Series of notebook computers. Cisco uses PRISM in its Aironet® series of WLAN products. Sharp Corporation recently announced a wireless audio/visual digital transmission system for the home that incorporates Intersil's PRISM technology.

Agere (Lucent)

As a key participant in the 802 LAN/MAN Committee and a founding member of WECA, Lucent was one of the first companies to produce 802.11 equipment and is involved in all phases of 802.11 from semiconductor manufacturing to end-user products. Its first 802.11b products sold under the WaveLAN® name. In early 2000, Agere (as Lucent) began marketing WLAN products under the ORiNOCO® brand. In March 2001, Lucent spun off its microelectronics division as Agere. The spin-off not only included the semiconductor business, but it also included the ORiNOCO product line. These products include NICs, APs, gateways, and outdoor 2.4 GHz systems and are discussed in the following section. In addition to its own products, Agere semiconductors are included in products sold through a variety of *original equipment manufacturer* (OEM) and reseller channels. These include Avaya (Lucent's enterprise communications spin-off), Apple Computer, Dell Computer, Hewlett-Packard, Gateway Computer, IBM, Compaq Computer, and Toshiba Corp. In May 2002, Agere announced the availability of 802.11a radio products for its dual mode radio access point.

Lucent Ventures has invested in Sychip, who produces chips for the wireless Internet appliance market, including WLANs. Employing IC design technology developed at Bell Labs, they produce single-chip solutions integrating radio and baseband functions.

Texas Instruments

A leading supplier of DSP, analog, and mixed signal ICs, Texas Instruments is also a very active member of the IEEE 802.11 LAN/MAN Committee. They produce a single-chip MAC/baseband controller with PCI, CardBus, and USB host interfaces for 802.11b systems. This controller supports the full range of 802.11b speeds up to 11 Mbps. In addition, it supports the optional 802.11b *packet binary convolutional coding* (PBCC™) modulation scheme for 22 Mbps operation. As mentioned in the discussion of 802.11g in Chapter 2, "The IEEE 802.11 Alphabet Spelled Out," Texas Instruments acquired the rights to this technology with its $300,000,000 acquisition of Alantro Communications Corporation in June 2000. PBCC is an optional method for achieving the higher data rates required by 802.11g.

Atheros

Atheros is a relative newcomer to the 802.11 arena. Dr. Teresa H. Meng of Stanford University founded it in 1998 to produce 5 GHz 802.11a chipset solutions. Privately held, its investors include a variety of venture capital firms as well as the corporate investors Proxim and TDK. In its short existence, Atheros has become one of the most visible and active proponents of 802.11a technology. Its first product consists of a single-chip transceiver and a single-chip MAC/baseband processor. The transceiver combines a power amplifier, a low-noise amplifier, and RF signaling components, supporting frequencies for IEEE 802.11a 5 GHz operation as well as the 5 GHz frequencies specified by *Mobile Multimedia Access Communication* (MMAC) in Japan. It supports all the 802.11a data rates up to 54 Mbps, but has an additional proprietary channel aggregation scheme that supports data rates up to 72 Mbps. The MAC/baseband processor provides 802.11a MAC functions, OFDM baseband processing, analog/digital conversion, and host interfaces (CardBus/PCI/MiniPCI). The chipset is completely CMOS based. CMOS is a well-known, cost-effective semiconductor technology that is more conducive to lower production costs than the more exotic compound semiconductor materials typically used for high-frequency radio applications. In September 2001, Atheros announced volume shipments at $35 per chipset for large-volume OEMs. These include Sony, D-Link, Intel, Proxim, NETGEAR, Actiontec, Intermec, and TDK.

Atheros also announced that Sony has integrated its chipset into WLAN cards and APs that are marketed for Sony's personal computing products. Sony also announced that it intends to incorporate 802.11a technology into its consumer electronics. In September 2001, Intel announced that it would be using the Atheros chipset in its family of 802.11a equipment that is expected to ship in November 2001. In October 2001, Proxim announced the availability of 802.11a networking products based on Atheros' chipset. See the sections on Intel and Proxim later in the next chapter for details. In addition to its IEEE 802.11a products, Atheros started sampling of an IEEE 802.11a/b/g 3-chip solution in March of 2002.

IceFyre

Founded in early 2001, IceFyre is a Canadian fabless semiconductor manufacturer focusing on the power consumption issues associated with 802.11a OFDM processing. Power consumption is a major consideration when designing portable devices and could inhibit the adoption of 802.11a if not addressed effectively. Seeking to meet this need, IceFyre has announced the development of a CMOS-based OFDM modem that provides substantial power savings for 5 GHz radio operation. IceFyre's solution includes radio, PHY, and MAC processing at a power consumption level that it claims is four times better than other OFDM processors while maintaining a 54 Mbps transmission rate over 100 meters. Like Atheros, its solution is CMOS based, but IceFyre's is a single-chip solution. IceFyre's design is based on DSP techniques that optimize the OFDM signal for more efficient processing by the power amplifier. This optimization not only enables more efficient use of the power amplifier, but it also reduces the need for several other external components that would add to the cost of an 802.11a chipset. IceFyre has said that they expect to sample a two-chip solution (an RF/PHY chip and a MAC chip) in mid-2002. An integrated RF/PHY/MAC chip is expected later that year.

Radiata (Cisco)

Radiata was founded in 1997 and acquired by Cisco in late 2000. It produces 5 GHz chipsets for the 802.11a market. Although Cisco uses Intersil's chipsets for its 802.11b products and has announced that it intends to continue that relationship for the foreseeable future, Cisco's 802.11a products will utilize Radiata's chipset. This chipset consists of a 5 GHz transceiver chip and a baseband modem chip. Like Atheros' offering, Radiata's chipset is completely CMOS based, helping keep production costs low.

Envara

Envara is a fabless chipset maker based in Israel that has announced some very ambitious plans for the 802.11 chip market. Although primarily focused on the 5 GHz 802.11a chipset market, Envara offers a chipset designed for backward compatibility with 2.4 GHz 802.11b systems. Founded in mid-2000 as hLan, Envara is positioning its products to operate with the large installed base of 802.11b equipment while taking advantage of the fast growing 802.11a market. It has announced a two-chip architecture incorporating a CMOS baseband chip with integrated MAC and 802.11a/b modem functionality. The second chip is an SiGe dual-mode radio with 2.4 and 5 GHz capabilities. Envara has announced that the chipset will be software upgradeable to comply with 802.11e (QoS), 802.11h (*dynamic frequency selection* [DFS] and power control), and 802.11i (security). The chipset will require external power amplifiers. Envara has submitted patent applications for a part of the implementation that produces increased transceiver sensitivity of 2 to 3 dB and reduced power consumption of up to 30 percent. It has announced that single-mode 5 GHz chipsets and reference designs for dual mode will ship in 2002.

Embedded Wireless Devices

Embedded Wireless Devices is another fabless semiconductor manufacturer who is seeking to take advantage of the coexistence of 2.4

and 5 GHz wireless networks. They have announced two software-programmable baseband processor products. Each supports concurrent operation on both the 2.4 and 5 GHz bands. The addition of an external radio interface(s) enables the support of 802.11a, HIPERLAN/2, 802.11b, Bluetooth, and *Home Radio Frequency* (HomeRF). Utilizing a proprietary embedded operating system, both products combine a 32-bit RISC processor, a 32-bit DSP, and various memory modules that provide baseband switching, I/O, and an interface to radio and external processors.

Cirrus Logic (ShareWave)

Best known for its DSP and analog ICs, Cirrus Logic jumped into the 802.11 chip market with its acquisition of ShareWave for $92 million in July 2001. ShareWave is focused on developing MAC processors for the wireless networking market. ShareWave does not provide radio modem chips; rather, it provides reference designs for the radio interface and is best known for its proprietary Whitecap™ technology. Whitecap and its successor Whitecap2™ are designed to operate over 802.11 WLANs and provide QoS functionality for multimedia and other QoS-based applications. It includes *forward error correction* (FEC), dynamic streams, and channel agility. As you may recall, these are some of the capabilities that the 802.11e Working Group is standardizing as part of 802.11's family of standards, allowing QoS applications over 802.11a and 802.11b networks. Although Whitecap is a proprietary technology, ShareWave is a very active member of the 802.11e Working Group, and the 802.11e specification will most likely include elements that are very similar to Whitecap. Share-Wave has indicated that Whitecap devices will be software upgradeable to 802.11e when it is approved.

Multimedia support is one of the key enablers for the adoption of 802.11 as a wireless networking technology beyond home computers. As we have seen, consumer electronics companies have started to incorporate 802.11 capabilities into home audio and video equipment, including set-top boxes, televisions, video recorders, and a variety of other devices. Cirrus is well positioned to capitalize on this market. It is currently shipping several versions of its basic MAC controller, including PCI, USB, PCMCIA, and MiniPCI interfaces. These controllers are already included in products from NETGEAR and Panasonic.

Raytheon

Raytheon was one of the founding members of the WLANA and manufactures 802.11 *frequency-hopping spread spectrum* (FHSS) consumer products and components that are marketed by Raylink. Raytheon entered the 5 GHz radio chip market in February 2001 with the introduction of its Tondelayo™ chipset. Paying homage to Hedy Lamar (a holder of the original spread spectrum patent), it is named after one of her most famous characters. The four-chip product includes a baseband IC, an I/F IC, a frequency converter/low-noise amplifier, and a power amplifier switch. Unlike other vendors seeking to utilize a single semiconductor technology such as CMOS, Raytheon has chosen to use three different materials for the chipset. Using CMOS materials, the baseband IC provides 802.11a PHY and MAC functionality. The I/F IC and the low-noise amplifier are fabricated with SiGe. Designed to accommodate the demands of OFDM processing, the power amplifier/switch module includes an amplifier made with Raytheon's proprietary PHEMT transistor process and two high-speed switches in a module based on *gallium arsenide* (GaAs). In late 2001 Raytheon Commercial Electronics began a collaboration with Systemonics of Germany.

Based in Dresden, Germany, and San Jose, California, Systemonic was founded in 1999 and specializes in flexible DSP solutions. In July 2001, they announced the availability of a multiradio baseband processor designed to support both 802.11a and HIPERLAN/2. It is a single-chip CMOS solution. They also announced the availability of a reference design platform that includes 5 GHz RF, PC Card and PCI interfaces, memory, and a CMOS baseband processor.

Mobilian

Mobilian was founded in 1999 to address the need for 2.4 GHz radio solutions that would span both Bluetooth and 802.11 applications. It is funded by venture capitalists, which include Vitesse and Dell Ventures. As 802.11b and Bluetooth have come to be viewed as complementary rather than competing technologies, a market is emerging for products that may incorporate both technologies. Mobilian has announced a two-chip solution that consists of a multistandard 2.4 GHz analog IC and a digital IC for baseband and MAC functionality. The radio modem chip uses a dual-antenna approach. Each antenna either transmits or receives for both Bluetooth and 802.11b. The radio chip has separate transmit and receive paths for each standard. For example, there are separate power amplifiers for the outbound Bluetooth and 802.11b signals, and inbound signals are passed to separate low-noise amplifiers. Firmware serves as a traffic cop by processing inbound signals and directing them to the correct receive path. The digital IC similarly has separate MAC and baseband processing paths for each standard—again with the firmware directing things. Mobilian claims that an integrated solution will have cost and packaging benefits over implementations with discrete chipsets for each standard.

Resonext

Originally founded as NeoSilicon in January 2000, it changed its name to Resonext in late 2000. Resonext is a fabless semiconductor company developing CMOS-based 5 GHz chipsets for the 802.11a and HIPERLAN/2 markets. In July 2001, they announced a partnership with IMEC, a Belgian research organization, for the use of IMEC's OFDM technology in Resonext products. In August 2001, they announced a zero I/F architecture for 802.11a that converts 5 GHz signals directly to baseband. As with all zero I/F architectures, the goal is fewer components, lower power consumption, and reduced cost. Resonext's two-chip solution consists of a radio modem chip and a baseband/MAC processor. It is marketed in two versions: a client version and an AP version. The two versions share the same radio chip, but have unique baseband/MAC chips. The client version supports PC Card, PCI, and MiniPCI interfaces. The AP chipset has an RISC processor and an 802.3 MAC interface. Resonext has indicated that both chipsets will have native support for 802.11e (QoS), 802.11i (including 802.1x, *Advanced Encryption Standard* [AES], and RC4), and 802.11h for European operation. Resonext has announced quantity pricing of the client chipset at $35 and the AP chipset at $45.

Other 802.11 Chipmakers

Magis Networks was founded in 1999 to produce chips and chipsets for the 5 GHz 802.11a and HIPERLAN/2 markets. Its efforts are focused on home networking and wireless multimedia distribution. They have indicated that their products will support both 802.11a and HIPERLAN/2 radio in front of a subset of the HIPERLAN/2 MAC.

Athena Semiconductors was launched in April of 2002 through the acquisition of wireless technologies of Comsilica and Hellenic Semiconductor Applications of Greece. Athena is developing a line of 802.11a WLAN chipsets.

Atmel has been producing 802.11 MAC products since 1999 and joined WECA in March 2000. They produce 802.11b MAC chips with PCMCIA and USB interfaces and a wireless-to-Ethernet bridge product that is used by AP manufacturers such as Linksys and NETGEAR.

Philips Semiconductors announced its second-generation 802.11b radio IC in early 2001. With an eye toward reducing the cost, they employ a zero I/F architecture that reduces the number of components on the chip. For semiconductor radio components, RF Micro Devices offers a variety of power amplifiers, low-noise amplifiers, mixers, and transceivers that can be used to build 802.11 devices.

Programmable Chips—SoC, FPGA, and SDR

The chipsets discussed so far have been various forms of fixed-function communications processors, but some participants in the 802.11 chip field have taken a different approach by developing programmable chips that are capable of supporting different standards.

Tality Corporation was spun off from Cadence Design Systems in late 2000 and is the former electronic design services group of

Cadence. They develop and design semiconductor solutions for a number of technologies. In April of 2002, Tality demonstrated a FPGA baseband/MAC controller designed to support IEEE 802.11a and IEEE 802.11b. The controller has been licensed to several customers.

BOPS, Inc. develops, designs, and licenses intellectual property for SoC DSP designs. In November 2001, they announced the availability of an evaluation chip providing support for multiple WLAN PHY interfaces. The programmable DSP core supports both 2.4 and 5 GHz networks, including 802.11b, 802.11a, HIPERLAN/2, and Japan's MMAC. BOPS, Inc. works with Tality and others to offer SoC DSP cores as well as development tools and design services for the wireless chip market.

Xilinx is one of the leading suppliers of FPGA processors and their products are used in a variety of wireless applications. One user of Xilinx FPGA processors is M-Tec wireless. M-Tec has implemented a HIPERLAN/2 MAC and also offers 5 GHz radio products supporting IEEE 802.11a, HIPERLAN/2, and HiSWAN.

The coexistence of multiple WLAN standards has created an opportunity for programmable processors. These products provide flexibility to equipment manfacturers producing multistandard products for the WLAN marketplace.

Equipment Makers and Carriers

As we have seen, the *Wireless Ethernet Compatibility Alliance* (WECA) has certified over 300 IEEE 802.11b products from over 70 vendors, including *network interface cards* (NICs), *access points* (APs), gateways, bridges, and outdoor networking systems. In addition, an increasing amount of IEEE 802.11b equipment is marketed with *personal computers* (PCs), computer peripherals, print servers, mobile devices such as laptops or *personal digital assistants* (PDAs), and consumer electronics products. Although each vendor's equipment is packaged differently, Figures 5-1 through 5-3 show some typical equipment from Symbol Technologies.

Figure 5-1
Symbol
Technologies AP

Figure 5-2
Symbol
Technologies
PC card

Figure 5-3
Symbol
Technologies
Peripheral
Component
Interconnect
(PCI) card

Figure 5-3
Symbol
Technologies
Peripheral
Component
Interconnect
(PCI) card

As the IEEE 802.11b market is maturing, the IEEE 802.11a equipment market has started to see the release of 5 GHz IEEE 802.11a products. The availability of chipsets has enabled the release of IEEE 802.11a products from Lucent, Proxim, Intel, Actiontec, Intermec, and SMC. The ratification of the IEEE 802.11g draft standard has prompted Intermec, 2002 and 2003 are key years for IEEE 802.11b as equipment for the new higher-speed standards is positioned as an alternative to IEEE 802.11b in the business market.

There are also a growing number of service providers offering *wireless local area network* (WLAN) services. These include public carriers, *wireless Internet service providers* (WISPs), government-sponsored WLAN access areas, and even quasilegal underground networks called *parasitic grids*. In this chapter, we'll discuss some of the key products and services that are making IEEE 802.11 a reality in the marketplace.

IEEE 802.11 Component and System Vendors

Intel (Xircom)

Intel has historically segmented its WLAN products into home and business product lines. Until the first quarter of 2001, it had promoted *Home Radio Frequency* (HomeRF) as the preferred home networking technology and IEEE 802.11 as the preferred enterprise technology. HomeRF is a 2.4 GHz wireless networking technology that is incompatible with IEEE 802.11. As a member of the HomeRF Working Group, Intel joined with Motorola, Siemens, Proxim, and others to become one of the main proponents of HomeRF. In 1999, Intel took an equity interest in Proxim and began marketing a full line of HomeRF products. Designed for home networking, the 1.0 version of HomeRF included voice and multimedia support, but only supported data rates of 1.6 Mbps. During 2000, members of the HomeRF Working Group announced that they were developing HomeRF 2.0, which would support data rates up to 10 Mbps. In early 2001, HomeRF 2.0 had not yet been completed. IEEE 802.11b had a firm foothold in the 11 Mbps market and was gaining momentum.

Against this backdrop, Intel announced that it was going to support IEEE 802.11b as the underlying technology for its home networking products and continued to promote IEEE 802.11 technology as the basis for the enterprise market. Although Intel still sells a 1.6 Mbps HomeRF product as an entry-level home system under the AnyPoint name, it has not adopted HomeRF 2.0 (released in mid-2001) for its higher-speed products. Intel now offers an 11 Mbps IEEE 802.11b AnyPoint system for home use. The AnyPoint line offers an NIC starting at $129 and an AP/gateway at $259.

Intel's IEEE 802.11 products for the enterprise market are offered under the Intel PRO/Wireless name. These products include both 11 Mbps IEEE 802.11b and 54 Mbps IEEE 802.11a equipment. Its IEEE 802.11b products include NICs and APs. They offer IEEE 802.11a APs and NICs as well. For organizations with existing 2.4 GHz IEEE 802.11b networks, Intel offers an optional dual-mode upgrade kit for the AP that enables simultaneous operation with both IEEE 802.11a and IEEE 802.11b networks.

In early 2001, Intel expanded its presence in the IEEE 802.11 market with the acquisition of Xircom for almost $750 million. Xircom is a leading vendor of both wired and wireless NICs and equipment. Intel already had an equity interest in Xircom. They also had a longstanding relationship with Xircom from supplying PC cards sold under the Intel brand. Xircom's strong presence as a supplier of products designed for the PDA and mobile computing markets was a complement to Intel's existing WLAN product line. Xircom's product line includes a Type II PC card for mobile computing platforms as well as specialized products for PDAs. In the middle of 2001, Intel introduced IEEE 802.11b WLAN products for Handspring and Palm handhelds. Intel's SpringPort module for the Handspring Visor plugs into the expansion slot on the back of the handheld. Similarly, the module for the Palm m500 plugs into the back of the handheld. Interestingly enough, these two products utilize the IEEE 802.11b technology from Cisco. Xircom had a joint development agreement with Cisco for IEEE 802.11b products prior to being acquired by Intel.

As a leading semiconductor vendor, Intel has to this point used chipsets from other vendors such as Intersil. In 2002 Intel announced its first chipsets for IEEE 802.11a/b. It also announced future plans for integration of IEEE 802.11 technology into its other semiconductor products.

Intel's consolidation of products under the IEEE 802.11 umbrella and its membership on the board of directors of the WECA since September 2001 have cemented its position as a leading member of the IEEE 802.11 community.

Proxim

Proxim has been supplying wireless networking products in various forms since 1989. Because of its long history in the wireless networking industry, the evolution of its products provides a roadmap for the evolution of WLANs. In 1992, Proxim introduced its 900 MHz RangeLAN product and continued with the 2.4 GHz RangeLAN2 in

1994. Targeted to the corporate market, RangeLAN2 is a 1.6 Mbps frequency-hopping wireless network system that uses the OpenAir standard. OpenAir was a proprietary standard developed and promoted by Proxim prior to the creation of IEEE 802.11. In 1998, as the IEEE 802.11 standard became more popular, Proxim introduced RangeLAN802, a frequency-hopping IEEE 802.11 product. That same year, it also began marketing products for the home and small office market with the introduction of its 2.4 GHz Symphony line. Proxim's Symphony products are all 2.4 GHz frequency-hopping systems and come in three variations: Symphony Cordless, Symphony-HRF, and Symphony HomeRF. The Symphony Cordless networking products are based on the OpenAir standard, providing 1.6 Mbps operation. The Symphony-HRF products also operate at 1.6 Mbps in the 2.4 GHz band, but are based on the HomeRF 1.2 standard. Introduced in October 2001, the Symphony HomeRF products adhere to the HomeRF 2.0 standard and provide 10 Mbps.

In June 2000, Proxim acquired Farallon for approximately $14 million in stock and cash. Farallon's products included IEEE 802.11b NICs for Apple's Macintosh. These became the basis for Proxim's Skyline brand of IEEE 802.11b products. Skyline IEEE 802.11b products now include NICs, gateways, and APs for both Apple and PC systems. In September 2001, Proxim introduced an IEEE 802.11b broadband gateway product for $299 that includes a four-port Ethernet switch, a router, and an IEEE 802.11b AP. Its network interface options include a *universal serial bus* (USB) adapter as well as PCI and PC cards.

Proxim's latest group of products for WLANs is called Harmony. Introduced in October 2000, Proxim has positioned it as an architecture that encompasses a variety of wireless standards, including IEEE 802.11a, IEEE 802.11b, IEEE 802.11g, HIPERLAN/2, OpenAir, Bluetooth, and others. Currently, Proxim offers four groups of products in its Harmony product line: IEEE 802.11a APs and NICs, IEEE 802.11b APs and NICs, OpenAir APs and NICs, and infrastructure products.

Accelerating Proxim's entry into the IEEE 802.11a market was their acquisition of Card Access, Inc. for $4 million in August 2001. Card Access is a manufacturer of IEEE 802.11a boards and components, and its products will be assimilated into Proxim's IEEE

802.11a offering. Proxim's first shipments of IEEE 802.11a products include interface cards and software for creating ad hoc peer-to-peer networks. The IEEE 802.11a AP provides standard 54 Mbps operation, but it also operates in a proprietary mode providing 100 Mbps operation. Proxim's IEEE 802.11a products are based on the Atheros chipset. Proxim has also added 802.1x security to its IEEE 802.11a products.

The Harmony IEEE 802.11b offerings include NICs and an AP. Infrastructure products include a power distribution system that supplies DC power to Harmony APs via an Ethernet cable. This reduces the need to run power to APs that already have Ethernet network connectivity. The Harmony architecture also includes an AP controller. Connected to APs via a standard Ethernet cable, the AP controller provides the central control and management of Harmony WLAN APs. This enables functions such as security and roaming to be managed across an infrastructure *basic service set* (BSS), regardless of the radio technology used by the AP.

Any discussion of Proxim would not be complete without a mention of the lawsuits it has filed regarding its patents for IEEE 802.11 technology. On March 8, 2001, Proxim filed a patent infringement lawsuit against Intersil, 3Com, Symbol, Cisco, Standard Microsystems, and Wayport. It claims that these companies have violated its patents and wants to collect royalty fees for products incorporating that technology. The lawsuit has started a wave of countersuits and even Agere, who was not named in the suit, has now sued Proxim for infringement of Agere patents. With the maze of patents held by the participants, the lawsuit may drag on for years, but it does not appear to have slowed the rollout of IEEE 802.11 products.

Coincidentally, while the lawsuit was filed, on March 21, 2001, Intel announced that it was refocusing its wireless home networking strategy to use IEEE 802.11b. This was notable because Intel has an equity interest in Proxim. Together they had been two major backers of HomeRF products. Intel's announcement caused an almost 60 percent drop in Proxim's stock price because Intel accounted for between 10 and 15 percent of Proxim's sales at that point in time. Although partnered with Proxim for HomeRF, Intel uses Intersil as the supplier for its IEEE 802.11b chips and also has an equity interest in Symbol. Both Symbol and Intersil were named in the lawsuit.

Proxim remains a major participant in the WLAN market and continues to produce a wide range of products.

Cisco (Aironet)

Cisco's entry into the IEEE 802.11b equipment market was jump-started by its acquisition of Aironet Wireless Communications. It acquired Aironet for $800 million in November 1999. Founded in 1993, Aironet produced IEEE 802.11b NICs, APs, and bridges and was one of the founding members of WECA. In March 2000, Cisco introduced the Cisco Aironet 340 series of products. It quickly became one of the best selling IEEE 802.11b product lines. Utilizing Intersil's PRISM® chipset, the Aironet series continues to be one of the leaders in the IEEE 802.11 marketplace. In January 2001, Cisco introduced the Aironet 350 series. Cisco offers NICs, APs, and bridges for IEEE 802.11b networks. It also offers an Ethernet-line-powered solution for its APs. Cisco has responded to IEEE 802.11b security concerns by offering a full range of 128-bit *Wired Equivalency Protocol* (WEP) systems and 802.1x user authentication through the Cisco Access Control server. With the acquisition of IEEE 802.11a chipmaker Radiata and its leadership in *orthogonal frequency division multiplexing* (OFDM), Cisco has introduced a dual mode (IEEE 802.11a/b) access point called the Aironet 1200.

Agere (Lucent)

Agere is not only a leading chipmaker, but it also offers ORiNOCO™, one of the first and most well-known brands of IEEE 802.11b equipment. As mentioned before, its first IEEE 802.11b products sold under the name WaveLAN™. This has been superceded by the ORiNOCO product line, which includes NICs, APs, gateways, Ethernet power distribution systems, and outdoor 2.4 GHz systems. Its network interface products offer both 64 and 128-bit WEP encryption. The AP-2000 is its latest AP offering. Responding to recent market concerns, the AP-2000 incorporates features that address the need for IEEE 802.11a support and enhanced security. It has two radio card slots that enable IEEE 802.11b and IEEE 802.11a cards to operate simultaneously in the same AP. This will facilitate the migration of IEEE 802.11b LANs to IEEE 802.11a. Agere does not currently offer IEEE 802.11a NICs. Enhanced security is provided by its support of 802.1x. Designed to interoperate with the Windows XP 802.1x clients, the AP-2000 utilizes certificate-based user authentication through a *Remote Access Dial-In User Service* (RADIUS) server and bidirectional automatic key distribution. The AP-2000 also provides wireless bridging capabilities.

Agere also offers a line of outdoor IEEE 802.11b products for longer-distance point-to-point or point-to-multipoint networks. The COR-1100 is a central IEEE 802.11b router and the ROR-1000 is a remote transceiver used in point-to-multipoint systems. These outdoor systems utilize a variety of antennae for both *Federal Communications Commission* (FCC) and *European Telecommunications Standards Institute* (ETSI) operation. They are capable of 11 Mbps operation at distances of up to 19 km (12 miles). A typical point-to-point application is a building-to-building wireless bridge that utilizes two COR-1100 routers. A typical point-to-multipoint application is a campus distribution system that uses a COR-1100 with multiple ROR-110 units. This system provides support for managed bandwidth connections and *virtual LAN* (VLAN) networks.

Symbol Technologies and Intermec Technologies

Founded in 1975, Symbol Technologies has been an active participant in the definition and application of WLAN technology in the industrial marketplace. An industry-leading provider of barcode-scanning systems, Symbol has leveraged its expertise in mobile information appliances with WLAN technology to create products for a variety of industries. Symbol claims over 70,000 installed WLAN sites. Its industrial products integrate WLAN technology with scanners, point-of-sale devices, barcode scanners, and other terminal equipment. For the consumer market, it offers IEEE 802.11 frequency-hopping systems for low-speed applications, and IEEE 802.11b and IEEE 802.11a systems for higher-speed applications. In addition, it supplies *original equipment manufacturer* (OEM) components and development packages for WLAN equipment manufacturers. Symbol's Spectrum24 series of products includes the AP 41X1 series of APs as well as PCI, PC Card, and Compact Flash network adapters. It also offers the BIAS-T connector, which supplies DC power over Ethernet cable. Optional 128-bit encryption and support for Mobile IP are also offered.

One of the most interesting products from Symbol is its NetVision phone shown in Figure 5-4. This device combines a thin-client web browser and a *voice over IP* (VoIP) telephone in a device similar to a cellular phone. It supports the VoIP protocol H.323 and enables simultaneous voice and data sessions. Integration with standard VoIP gateway equipment from Ericsson, Nortel, Cisco, and others allows both in-building calls and provides access to the *Public Switched Telephone Network* (PSTN).

Another manufacturer with a focus on the industrial market is Intermec Technologies. One of the pioneers in barcode scanning, its products include *radio frequency ID* (RFID) systems, remote data acquisition, and WLANs. Like Proxim, Intermec offers products based on the OpenAir standard. Intermec also supports IEEE

Figure 5-4
Symbol
Technologies
WLAN phone

802.11b and IEEE 802.11a in its MobileLAN product family. The MobileLAN product group includes Ethernet power equipment, APs, PC Cards, management software, and a voice solution. The voice solution includes H.323 handsets, an 802.11b gateway supporting up to 32 simultaneous calls, and interfaces to standard *private branch exchanges* (PBXs) from manufacturers such as Nortel, Lucent, and Siemens.

Enterasys

Enterasys is the former enterprise networking business of Cabletron and became an independent company in August 2001. Enterasys offers complete indoor and outdoor systems under the RoamAbout product family. The indoor RoamAbout products include IEEE 802.11b APs and NICs. In addition, the RoamAbout R2 is designed to house two different radio cards to facilitate migration from today's IEEE 802.11b to IEEE 802.11a. Enterasys outdoor systems provide long-distance IEEE 802.11b connectivity. Enterasys offers a 14 dB antenna that provides 11 Mbps at a distance of 5.6 km and a 24 dB antenna that enables distances of up to 40 km. The use of an omni-directional antenna allows point-to-multipoint operation. A typical application would be the interconnection of a set of buildings to a hub in a campus network.

Microsoft

In light of the major computer vendors' support for IEEE 802.11b wireless networking, it's not surprising that Microsoft would incorporate support for IEEE 802.11b into its operating systems. In the first quarter of 2001, around the same time that Intel refocused its WLAN strategy around the IEEE 802.11 WLAN family, Microsoft announced that it would incorporate support for IEEE 802.11b into its Windows XP operating system. This was a setback for Bluetooth proponents who had been hoping for Bluetooth support by Microsoft.

At the time, Bluetooth equipment was still experiencing growing pains and IEEE 802.11b had already proved to be commercially viable. Bill Gates was quoted as saying, "We thought it would take several years for the 802.11b format to catch on; we've been incredibly surprised by how quickly it has taken off." Although Microsoft may incorporate Bluetooth support in a future release, it has embraced IEEE 802.11b as a key part of its strategy for wireless computing.

Window XP's support for IEEE 802.11b includes automatic network detection and configuration. This enables a traveler to turn on a Windows XP device in an airport or hotel and automatically attach to the local IEEE 802.11b WLAN. In addition, Windows XP's support for IEEE 802.1x security has helped address some of the security concerns with IEEE 802.11b. Microsoft has also added wireless support in its *Network Driver Interface Specification* (NDIS). NDIS is important in that it enables wireless hardware devices to communicate with protocol stack software for network performance optimization and other protocol-specific features. It also enables you to control power usage and signal strength from an application. IEEE 802.11b equipment vendors are now able to certify products through the *Windows Hardware Quality Labs* (WHQL) process.

Demonstrating its commitment to IEEE 802.11b, Microsoft has installed the world's largest corporate IEEE 802.11b network at its corporate headquarters in Redmond, Washington, and over 50 locations throughout the Seattle area. There are over 3,000 APs throughout Microsoft's facilities and several thousand employees have high-speed wireless access to the corporate intranet as they roam between buildings. Microsoft has estimated a productivity gain of 30 minutes per employee per day. Multiplied by the several thousand users, this translates into substantial savings and a quick payback on the equipment. Component vendors are also supporting IEEE 802.11b functionality in Windows XP. For example, Intersil is providing drivers for components using Intersil's PRISM technology and has set up source code access and licensing agreements with IEEE 802.11b equipment OEMs. The driver supports 802.1x in Windows XP and version 5.1 of Microsoft's NDIS.

Computer Makers

All the major computer vendors have initiatives in the IEEE 802.11b market. Every platform from servers to PDAs is now available with built-in IEEE 802.11b support or standard wireless connectivity option packages. Coupled with IEEE 802.11b support by Microsoft, a whole new generation of wireless mobile computing devices has been born. In addition to selling hardware, computer makers like Compaq and IBM have also created extensive service organizations that design WLAN solutions and manage WLAN networks. Wireless data access has opened up a whole new universe of computing applications and computer makers are partnering with software providers to deliver industry-specific solutions from retail to education.

Apple Computer is one of the innovators in offering WLAN products as part of a computer product line. They have been marketing WLAN connectivity features for several years under the AirPort name and have a long-standing relationship with Agere (ORiNOCO) for IEEE 802.11b components. Leveraging their strengths in educational software, digital imaging, and wireless networking, Apple has become the leading provider of wireless computing equipment to the educational market.

In mid-2001, IBM announced a major commitment to wireless networking across all of its products and services. This included integrated IEEE 802.11b for all IBM ThinkPad and WorkPad devices (including an antenna built into the screen). In addition, IBM Global Services introduced design, installation, and network management services that can be bundled with IBM IEEE 802.11b products for turnkey networks. A third major investment was the establishment of programs to promote wireless product development. This included wireless training, support, and marketing initiatives with tens of thousands of IBM business and application development partners as well as independent software vendors. IBM's strategy is based on the belief that the adoption of wireless technology will accelerate rapidly in 2002 and 2003 due in large part to the productivity gains resulting from wireless business applications. According to IBM, they have seen productivity gains from 30 to 50 percent in some application deployments. As evidence of their confidence in this trend, they have invested over \$1 billion in wireless technology in 2000 and 2001, and expect to have over 10,000 people trained in WLAN networks in 2002.

IBM has also joined Cisco's Mobile Office initiative. Cisco is providing infrastructure equipment and software with IBM adding systems integration, network installation, and management services. Reinforcing IBM's information about productivity increases, Cisco indicated that it conducted a business survey of over 300 businesses and has seen productivity gains averaging over 20 percent from increased network connection time alone. Symbol Technologies has also joined with IBM to deliver wireless solutions to the industrial marketplace. IBM will provide services, marketing, and distribution in conjunction with Symbol's wireless equipment offerings.

Compaq is also very active in the promotion of IEEE 802.11b under the MultiPort name. It remarkets Agere's ORiNOCO wireless network products and offers integrated IEEE 802.11b in its Evo notebook computers and iPAQ PDAs. The Evo notebook has an antenna mounted in the notebook panel and supports either IEEE 802.11b or Bluetooth. Compaq announced in 2001 that it was going to add a focus on professional services to its overall corporate strategy. Given the iPAQ's firm foothold in the enterprise PDA market, it is natural that they would develop a line of consulting services for WLAN networking. These services are aimed at identifying wireless iPAQ applications and then designing, implementing, and managing the networks that support them.

Public WLAN Networks

Filling the gap between short-range *wireless personal area networks* (WPANs) and *wide area networks* (WANs) like *third-generation* (3G) cellular, IEEE 802.11b WLANs have enabled a new breed of broadband service provider. These providers have developed several business models and structures including WISPs, government-sponsored public access areas, and even free community networks sometimes pejoratively called parasitic grids. The commercial WISP is by far the most prevalent of these service providers.

The WISP

The public's desire for wireless high-speed data access, the failure of low-speed or proprietary networks, and the slow rollout of wireless data services by established carriers have provided an opportunity for IEEE-802.11-based telecommunications services. Just as demand for Internet access created the *Internet service provider* (ISP), demand for wireless connectivity created the WISP. Current WISP business models take several forms, but share IEEE 802.11b as the primary connectivity option, with IEEE 802.11a on the horizon.

The first group contains those carriers establishing regional or national networks catering to the mobile businessperson in public *hot spots*. These hot spot service providers may be new independent WLAN carriers or WLAN operators that are part of established ISP, wireline, or cellular operators. A hot spot is an area in a public venue such as an airport, hotel, convention center, coffee shop, or library that serves as a wireless access hub to the Internet. Subscribers to a carrier's service can use a laptop, PDA, or other IEEE-802.11b-enabled device to connect to the Internet whenever he or she is in a hot spot coverage area. The use of *virtual private network* (VPN) systems and authentication technology such as RADIUS enable secure connections.

The second group contains local service providers selling wireless Internet access to small businesses and residences similar to the *Multichannel Multipoint Distribution Service* (MMDS). WISP services may be offered as part of a local wired ISP's service portfolio, but there are also specialized WISPs that offer wireless Internet access where wired access is unavailable or unfeasible. Coastal islands or rural areas are typical examples. In addition, several companies are beginning to market centralized billing, authentication, and subscriber management services that bridge the gap between different carrier hot spots. Still in its infancy, the WISP industry has been characterized by the search for viable business models, much like the early days of the ISP industry. It has already seen its share of bankruptcies, consolidations, and ambitious network launches. What remained constant for the WISP industry in the difficult business climate of 2001–2002 was a steady increase in demand and the proliferation of low-cost interface equipment. This can only help the survivors as they refine their business plans and technology offerings. Industry studies have projected that wireless Internet service revenues will approach $1 billion by the end of 2005.

Hot Spot Service Providers

WISPs create IEEE 802.11b hot spots for the mobile professional by installing WLAN APs in public areas. Business travelers are able to use their own IEEE 802.11b devices to access the Internet in an airport lounge, a hotel room, or a local coffee shop where the WISP has provided a network. WISPs connect that WLAN to the public Internet and/or a private data network, selling Internet service or remote corporate access in several ways. It may be sold to a user on a per-use basis, much like using a payphone. WISPs may sell prepaid services, much like using a prepaid telephone calling card. Rates at each usage location vary and the card is debited after each use. WISPs also sell subscriptions to their service on a monthly or yearly basis, much like using a cellular phone. Subscriptions are usually sold on an access time basis and unlimited access plans are available. Some WISPs may also charge a fee based on the amount of data transferred, and access location owners such as a hotel may add a surcharge. Connection to the network is generally made by setting the *Service Set Identifier* (SSID) of the user's equipment to the name of the hot spot network. When an Internet browser session is started, the local network's welcome screen appears for entry of subscriber login information, prepaid service information, or credit-card information for nonsubscribers.

The first wave of hot spot networks operated as independent entities and competed for end users. It soon became apparent that a primary driver for the adoption of this technology was the widespread availability of coverage. The time and costs of providing availability over a large area (even when concentrated in hot spots) are prohibitive for a single carrier. Subscribers are hesitant to buy until they have a wide range of available locations, and it's difficult to expand the network without a fast growing subscriber base to fund operations. This catch-22 has led to the creation of business models that are more like the cellular industry. Roaming agreements, central clearinghouse functions, and the ability to share subscriber profile information as customers move about fueled the cellular phone industry growth of the last 20 years. This model allows the ownership of a subscriber, but extends coverage through reciprocal access. The application of these functions to the developing WISP industry are key to its growth.

Although growth in the demand for wireless Internet access is continuing, there have been several bumps in the road for hot spot WISPs. The first problem is that network buildouts require significant up-front capital expenditures. If customer growth and service revenues do not ramp up fast enough, cash flow and debt service problems can result. Because the target market of the large WISPs is the business traveler, the difficult business climate and travel industry problems of 2001–2002 have dampened demand. This has exposed some of the weaker business plans and caused some high-profile failures and corporate restructurings. The good news is that even in this environment, demand is continuing to grow. This has been validated by the entry of established service providers into the public hot spot market.

The second set of problems involves the immaturity of back-office operations, including billing, subscriber account management, and the lack of intercarrier roaming agreements. In particular, the lack of intercarrier roaming agreements is similar to the early days of the cellular phone industry, when a subscriber's phone would only work in its home area. Without widespread roaming agreements to supplement network buildouts, the hot spot WISP subscriber will be subject to spotty coverage. Requiring separate accounts for widespread coverage is not acceptable to the consumer.

WECA is examining these issues in a Working Group called *Wireless ISP Roaming* (WISPr). Members of WECA's WISPr Working Group include Dell, Agere, GRIC, hereUare Communications, Nortel, Nokia, Wayport, Enterasys, and others. They are working to establish a set of recommended practices for roaming and coordinated billing. An example of this would be the use of standard service provider IDs as part of a user's account login. That ID and standardized information about the subscriber's session could be passed to a billing clearinghouse for distribution to the service provider that owns the account. The service provider would then be able to compensate the owner of the hot spot according to the terms of a roaming agreement. The subscriber would then receive a single bill with the details of session usage, roaming charges, and so on. Recommendations are expected sometime in 2002. Companies like hereUare Communications and Excilan are developing infrastructure products and services in support of intercarrier cooperation.

In the meantime, the demand for global WLAN roaming services is being addressed by established remote access providers (such as GRIC and iPass) and *Global System for Mobile Communications* (GSM) operators (such as Sonera, Telia, and Jippii Group Oyj). Remote access service providers GRIC and iPass have supplemented their existing dial-up roaming services by forging alliances with IEEE 802.11b hot spot owners such as Wayport and Concourse. In Europe, Jippii Group Oyj, Sonera, and Telia have IEEE 802.11b WLAN hot spot operations. The remote access providers typically market their remote connectivity services to corporations. They have historically provided dial-up and *Integrated Services Digital Network* (ISDN) services to the business traveler. By forging agreements with IEEE 802.11b WISPs, they can now offer high-speed wireless access as well. This enables the WISP to sell service to a large established customer base with authentication, billing, and subscriber management operations provided by the remote access partner. The trade-off is that they cannot own the subscriber. The remote access operator does not have the cost of building the hot spot network and is now able to market new products based on high-speed wireless access. Revenues are shared in some manner.

GSM operators also have established billing and subscriber management operations. In addition, GSM operators have an extensive set of roaming agreements between operators worldwide. Some GSM operators have leveraged this existing infrastructure, created their own hot spot networks, and marketed WLAN access services to the GSM subscriber. Nokia has promoted this effort by creating equipment that enables you to authenticate and register a WLAN user through the GSM infrastructure. In GSM, a *Subscriber Identification Module* (SIM) card that is installed in a cell phone provides services and authentication. Nokia has combined a SIM card with IEEE

802.11b, enabling travelers to use hot spots wherever there is a roaming agreement in place between GSM operators. This is a huge potential market. Nokia estimates that there are 500 million roaming calls made every month on GSM networks and 20,000 roaming agreements across the globe. This potential market of mobile customers and the relatively low cost of installing IEEE wireless access has prompted some GSM operators to see WLAN hot spot services as a shortcut to 3G data revenues.

With WISPs becoming more like cellular carriers and cellular carriers rolling out wireless data services, it's understandable that the relationship between WISPs and cellular operators has become somewhat controversial. On one side are those who believe that WISPs and WLANs will become the local wireless data access method of choice, stifling the adoption of 3G. On the other side are those who believe that WLANs will remain a niche implementation and that the primary wireless data delivery mechanism will be 3G and *fourth-generation* (4G) cellular technologies. The middle ground sees both technologies addressing specific needs and coexisting as complementary solutions. This is the more likely scenario. 3G and WLAN are different in many respects. WLANs are optimized for high-speed localized operation and are well suited for data-intensive applications. 3G cellular offers slower speeds, but wider coverage and greater mobility. Although there will be some overlap, these sets of characteristics will likely lead to very different applications using each technology. The larger problem is that the most profitable data revenues from each will most likely be derived from the corporate user. It's likely that both technologies will find their proper place in the market. The real question is who will own the subscriber.

Wayport

Founded in 1996, Wayport, Inc. has been providing high-speed Internet connectivity to the hotel industry since 1997. Since that time, Wayport has become a leading supplier of broadband access to hotels and airports around the world with installations in over 400 locations. Wayport's products and services provide wired and wireless IEEE 802.11b Ethernet access to rooms and common areas such as conference rooms and lobbies. Wayport's hotel and resort installations include Wyndham, Four Seasons, Hilton, Sheraton, Marriott, Westin, Radisson, and Doubletree. IEEE 802.11b airport coverage includes Austin, Dallas, Seattle, and San Jose. Wayport also offers airport business centers with wired broadband access in Atlanta, Chicago, New York (LaGuardia), and others. In August 2001, Wayport acquired the operations of AerZone from SoftNet, AerZone's parent company. AerZone operated wired LapTop Lanes airport business center operations and had contracts with United and Delta for IEEE 802.11b wireless access at the airlines' gates and lounges. Wayport's acquisition of LapTop Lanes expands its business center presence in several new airports and provides locations for future wireless expansion.

Another major partnership was announced in September 2001. Wayport has teamed with iPass to provide broadband access (including IEEE 802.11b) to iPass customers. This partnership enables Wayport to expand its network usage through exposure to large iPass corporate users, while iPass can offer its customers access at Wayport locations worldwide. Three factors have helped Wayport weather the economic challenges of 2001: technology diversification, a revenue sharing business model, and a diversification of its target market. Wayport's diversification in wireless and wired broadband access has enabled it to derive wired access revenues while the wireless access market ramps up. Revenue sharing agreements with its partners have enabled it to offload up-front network buildout costs. Although this results in less revenue per subscriber, it creates a larger customer base through partner marketing efforts and increases overall network usage. Wayport also offers different service packages targeted at large corporate users as well as individuals. Wayport offers individual memberships at $29.95 per month for a yearly contract, prepaid services in $49.95 increments, and corporate memberships based on the number of subscribers.

T-Mobile Broadband/VoiceStream/MobileStar

Another public hot spot carrier founded in 1996 is MobileStar. MobileStar is focused on the wireless access market and packages its services for the individual user. Like Wayport, MobileStar provides WLAN coverage in airports, hotels, and other venues frequented by business travelers. Its most notable partner is Starbucks. In early 2001, it announced that it would be installing IEEE 802.11b access in 4,000 Starbucks' locations by 2003 and by September 2001 had installations in 500 Starbucks locations. Partnering with Starbucks and MobileStar were Microsoft and Compaq who would be providing iPAQ Pocket PCs for in-store use. MobileStar has an agreement with American Airlines to provide WLAN service in American's Admirals Club executive airport lounges. MobileStar also provides general airport coverage to airports in Dallas, Austin, and Louisville. A legacy of its early start in the market is several hotel installations using the OpenAir standard.

Unfortunately, in October 2001, MobileStar announced that it could not find financing for continuing operations and that it would be shutting its doors. The shutdown was short-lived, and in early November 2001 MobileStar announced that it was resuming operations. With so many solid partnerships in place, it is not surprising that MobileStar's assets were acquired. It was subsequently announced that VoiceStream Wireless (a subsidiary of Deutsche Telekom) was purchasing the assets of MobileStar through the bankruptcy court and had arranged financing for MobileStar to continue operations. VoiceStream Wireless is one of the first GSM operators in the United States. In late 2001, it became one of the first carriers to rollout 2.5G cellular data services. Similar to Telia and other European GSM carriers that operate hot spot networks, VoiceStream can now immediately offer high-speed data service in areas where its cellular service may not be available and build a presence for cellular service expansion. MobileStar Services have been rebranded as T-Mobile broadband. It offers prepaid service, as well as local and national subscription plans. Advertised prices list prepaid service for $50.00 for 300 minutes, unlimited local service for $29.95 a month, unlimited national service for $49.95 a month, and per-use charges of $2.99 per 15 minutes and $0.25 for additional minutes. There are also several variations offering monthly minutes plus a fee for additional minutes.

Other Hot Spot Networks

Surf and Sip is a hot spot carrier providing IEEE 802.11b access in restaurants, hotels, cafes, and other public areas. Surf and Sip enables a subscriber to use his or her own IEEE 802.11b equipment, but also provides on-site Apple computers for users without access devices. Surf and Sip currently has more than 60 locations across the United States. Surf and Sip also offers products and services that enable a business to serve as its own mini-WISP. This product manages broadband access on site (security, roaming, and so on), enabling the business owner to offer Internet access directly to the user.

Concourse Communications is a telecom services company providing IEEE 802.11b services to airports in Minneapolis (St. Paul) and Detroit. It has also formed a subsidiary called New York Telecom Partners, which contracted with the Port Authority of New York to provide general IEEE 802.11b services to New York Port Authority facilities in the fall of 2001. This includes the three New York area airports—LaGuardia, Newark, and Kennedy International. These airports currently have MobileStar IEEE 802.11b service in the American Airlines executive clubs. It also includes the Holland and Lincoln Tunnels, the Port Authority Bus Terminal, and the PATH public transportation system. One of the major investors in Concourse is SpectraSite Communications, one of the largest wireless tower operators in the world.

WiFi Metro is a California-based operator of IEEE 802.11b networks. It is backed by the same set of investors as hereUare Communications. Initial installations are centered in the San Francisco Bay area through the acquisition of Airwave's hotel and coffee shop installations. With the same backers as hereUare, it's not surprising that they would also be partnered with hereUare for access to additional locations. WiFi Metro also uses hereUare's subscriber access and back-office products. Charges are about $20 for unlimited access.

SkyNetGlobal is the largest IEEE 802.11b WISP in Australia. It covers 16 airports and several major hotels in Australia and Asia. It has a roaming agreement with MobileStar for about the same number of North American locations. SkyNetGlobal has signed an agreement with GRIC for roaming services on GRIC's remote access network. It charges monthly rates as well as a data surcharge.

RoomLinx is a WISP that provides wired and wireless broadband access to hotels and convention centers in the United States and Canada. It has about 10 IEEE 802.11b installations including the Los Angeles Convention Center.

Roaming and Hot Spot Access Services

Boingo Wireless is the brainchild of Earthlink's founder Sky Dayton. Started in early 2001 and publicly launched in late 2001, Boingo lists Sprint PCS among its initial investors. The concept behind Boingo is to provide a simple user client for free that detects and connects to popular IEEE 802.11b WLAN carriers. This eliminates the need to set up IEEE 802.11b configuration parameters like the SSID when changing carriers and simplifies the connection process. It also manages user profiles and provides a VPN connection to Boingo's data network. Boingo has announced WISP relationships with Wayport, Surf and Sip, RoomLinx, Air2Lan, AirPath, and hereUare Communications. With these relationships in place, Boingo claims to have a presence in the top 20 airports in the United States as well as in over 750 commercial locations. Boingo offers a per-use plan at $7.95 per 24-hour period, a monthly subscription plan for ten 24-hour connections at $24.95, and an unlimited usage plan at $74.95.

With over 14,000 locations across the globe, iPass is one of the largest providers of remote access services. Agreements in 2001 with Wayport and Concourse Communications have extended its portfolio of dial-up, ISDN, and wired broadband connectivity services to include IEEE 802.11b. iPass provides roaming, secure connectivity, and consolidated billing services to its subscribers regardless of the access technology.

GRIC Communications is another leading provider of remote access services. Like other access providers, it provides a customer management infrastructure and a secure centralized access method across technologies and networks. Network access is provided through its network of over 300 ISP and service provider partners. Primarily known for its extensive dial-up access network of more than 15,000 locations in 150 countries, GRIC has moved aggressively into the IEEE 802.11b remote access market. In the second half of

2001, it signed several agreements with IEEE 802.11b providers in the Asia-Pacific region. In August 2001, GRIC announced agreements with Japanese ISPs and service resellers including *Internet Initiative Japan* (IIJ) and NTTPC Communications. China Netcom is supplying IEEE 802.11b access in China. In Australia, GRIC has partnered with Skynet Global. Overall, GRIC expects to have over 1,000 wireless APs in 2002.

hereUare is a firm providing back-office infrastructure such as billing, roaming services, subscriber management, and network access software specifically for IEEE 802.11b networks. As mentioned previously, the same investors have started WiFi Metro and the two companies have a close relationship. In addition, hereUare has an agreement with PayStar Corporation to provide IEEE 802.11b wireless access at PayStar's information kiosks. The first phase of this initiative provides IEEE 802.11b at over 300 of these kiosks in hotels, airports, and other commercial venues.

Cellular and Telephone Carrier WLAN Services

Sonera is the largest mobile operator in Finland with almost 2.5 million subscribers. It also has international operations in Eastern Europe and Turkey. Sonera's WLAN operations include 50 hot spot installations in Finland. In October 2001, Sonera conducted a successful WLAN roaming feasibility test using GSM technology. Sonera uses Nokia's GSM/WLAN cooperative platforms, and Nokia has indicated that over 20 operators are in trials with the same technology.

Telia is a Swedish GSM operator that has partnered with Symbol Technologies to offer WISP hot spot services. With over 300 WLAN locations, it is expanding rapidly into Norway, Denmark, and Finland. Telia is participating in a WLAN trial with Tenzing Communications and Scandinavian Airlines. Tenzing Communications specializes in broadband communications for the airline industry and serves as an ISP for in-flight Internet access. The trial will test wireless in-cabin data services working in conjunction with Telia hot spot locations. A roaming agreement between Telia and Tenzing will provide subscribers (to either carrier) IEEE 802.11b WLAN access, whether in an airport or the airplane cabin.

Wireless Network Services Oyj was spun off from the Finnish telecommunications carrier Jippii Group Oyj as a WISP in February 2001. In addition to public hot spots Wireless Network Services is developing outdoor public access networks. It expects to have 400 APs installed by the end of 2001, and 900 APs installed by the end of 2003. It also has plans for installations in London and Frankfurt followed by Spain and Italy. Wireless Network Services has partnered with Symbol Technologies and Compaq to provide GPRS-based WLAN roaming services as an alternative to 3G wireless data.

NTT Communications is a Japanese wireline service provider that is planning to launch IEEE 802.11b wireless data services in 2002. This service is currently in a free-trial stage in about 20 Tokyo restaurants and coffee shops. NTT Communications service is positioned as an alternative to *Digital Subscriber Line* (DSL) service at about half the cost. Although initial service is based in IEEE 802.11b, IEEE 802.11a and *IP version 6* (IPv6) are being considered for subsequent services.

Community Networks

The spirit of free access and information sharing that characterized *GNU's Not UNIX* (GNU), Linux open-source code, and the early days of the Internet have given rise to the creation of free IEEE 802.11b community networks. It works like this: Utilizing standard IEEE 802.11b products and sometimes homemade hardware projects such as antennae made from Pringles potato chip cans, an enterprising person sets up an AP in a window or on a rooftop. It is then connected to a broadband Internet connection such as a cable modem or DSL, and neighbors and passersby are invited to surf as they please. The freenet community likens this to allowing your neighbors to use electricity from extension cords placed in your yard. AP owners concentrated in a specific area will usually join together in a common interest users group such as the *Bay Area Wireless Users Group* (BAWUG). Using the Internet as a common forum, the groups will meet to share implementation and technical information. The vision of the freenet community is a widespread series of loosely knit small local networks that are run for free or only have maintenance costs.

The commercial ISPs take a somewhat different view due to third-party access restrictions sometimes written into service contracts—hence, the term parasitic grids. Commercial service providers claim that unauthorized access prevents them from managing their networks and detracts from the service quality of authorized users. So far IEEE 802.11b freenets have remained small enough to be largely ignored by the wireless broadband industry. Whether this movement will go the way of Napster and be stifled due to its success or go the way of Linux and spawn cheap, shared commercial media is yet to be determined. There are over 100 freenets worldwide and new ones are created daily. Table 5-1 lists some of these community networks and community network user groups.

Table 5-1		
Community freenets	BAWUG	San Francisco, CA
	Seattle Wireless	Seattle, WA
	NYCWireless	New York, NY
	Pdx Wireless	Portland, OR
	Consume	London
	NoCatNet	Sonoma County, CA
	Personal Telco	Portland, OR
	BC Wireless	Vancouver, BC
	Green Bay Professional Packet Radio	Green Bay, WI
	NoVA Wireless	Leesburg, VA
	Madrid Wireless	Madrid
	Xnet Wireless	Australia
	Wireless France	France
	RDU Wireless	Raleigh-Durham, NC

Table 5-1 cont.	Guerrilla.net	Boston, MA
Community freenets	Barcelona Wireless	Barcelona
	StockholmOpen.Net	Stockholm
	CU Wireless	Champaign-Urbana, IL
	DallasFreeNet	Dallas, TX
	Austin Wireless	Austin, TX
	St. Louis Wireless	St. Louis, MO
	Doral Wirelesss	Miami, FL
	AirShare.org	San Diego, CA
	Prenzel.net	Berlin
	IrishWAN	Ireland
	NZ Wireless	New Zealand
	Houston Wireless	Houston, TX

Physical Concepts and Architecture

The unique demands of wireless data transmission over radio presents several challenges. Recent advancements in digital signal processing and radio transmission technologies have served as the catalyst for IEEE 802.11. These include spread spectrum radio and advanced modulation. Let's look at these in more detail.

Spread Spectrum Radio

One of the basic technologies underlying the IEEE 802.11 series of standards is spread spectrum radio. The fundamental concept of spread spectrum radio is the use of a wider frequency bandwidth than that needed by the information that is transmitted. Using extra bandwidth would seem to be wasteful, but it actually results in several benefits including reduced vulnerability to jamming, less susceptibility to interference, and coexistence with narrowband transmissions. There are several spread spectrum techniques including time hopping, frequency modulation (chirping), *frequency-hopping spread spectrum* (FHSS), *direct sequence spread spectrum* (DSSS), and hybrids of these.

Both FHSS and DSSS are used in the IEEE 802.11 series of standards. FHSS uses the extra bandwidth to create a series of frequency channels, each of which is capable of carrying the information. Transmissions are spread across the spectrum by hopping from channel to channel in a pattern that is known to the sender and receiver. Figure 6-1 illustrates IEEE 802.11 frequency-hopping signals. DSSS uses the extra bandwidth to include patterns of special-purpose data in addition to the actual information being transmitted. This extra data increases the amount of bandwidth required for each information bit, effectively spreading the information across a wider band of spectrum.

FHSS and DSSS are not modulation techniques, but simply methods of distributing a radio signal across bandwidth. In addition to spreading the signal across a frequency band, spread spectrum systems modulate the signal. *Modulation* is the variation of a radio signal to convey information. The base signal is called the

Figure 6-1
FHSS signal

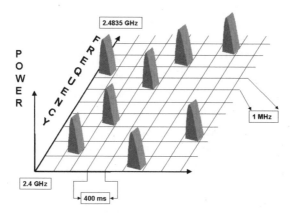

POWER

FREQUENCY

2.4835 GHz

1 MHz

2.4 GHz

400 ms

carrier. The variation may be based on the strength (amplitude modulation), frequency, or phase (frequency offset) of the signal. The modulation technique directly affects the data rate. Higher data rate modulations are generally more complex and expensive to implement. Modulations resulting in higher data rates pack more information in the same bandwidth. Small disruptions in the signal cause the degradation of more data. This means that the signal must have a higher *signal-to-noise ratio* (SNR) at the receiver to be effectively processed. Because a radio signal is stronger the closer it is to the source, the SNR decreases with distance. This is why higher-speed systems have less range. Examples of modulation techniques used in the IEEE 802.11 series of specifications are *binary phase-shift keying* (BPSK), *quadrature phase-shift keying* (QPSK), *Gaussian frequency-shift keying* (GFSK), and *complementary code keying* (CCK).

Whereas the creation of a spread spectrum transmission consists of spreading the signal and modulating the data, the reception consists of "unspreading" the signal (also called *correlation*) and demodulating the data. The key to correlating a signal is to synchronize the spreading code between the sender and receiver. In general, FHSS systems are more difficult to synchronize because they must synchronize both time and frequency. DSSS systems use a single frequency band and only have to synchronize the signal timing. This leads to lower latency when synchronizing the transmission.

FHSS

Spread spectrum radio techniques originated in the U.S. military in the 1940s. The unlikely co-patent holders on spread spectrum technology are the actress Hedy Lamar and musician George Antheil. Lamar had been married to a German arms dealer and fled Germany as the Nazis came to power. One of Antheil's techniques involved the use of player pianos. These two facts came together to create one of the twentieth century's most influential radio technologies.

The military had started to use radio as a remote control mechanism for torpedoes, but this technique suffered from a vulnerability to jamming. Aware of this, Lamar suggested to Antheil that the radio signal should be distributed randomly over time across a series of frequencies. The transmission on each frequency would be brief and make the aggregate less susceptible to interception or jamming. The problem was synchronizing the transmitter and receiver to the frequency being used at any point in time. Antheil used his musical expertise to design a synchronization mechanism using perforated paper rolls like those found in player pianos.

They were assigned patent number 2,292,387 and gave the rights to the Navy in support of the war effort. Although the Navy did not deploy the technology, engineers at Sylvania Electronic Systems applied electronic synchronization techniques to the concept in the late 1950s. The U.S. military began using these systems for secure communications in the early 1960s. The spread spectrum technique spawned from the work of Hedy Lamar and George Antheil is what we now call FHSS.

In FHSS, the information signal is overlaid on a narrowband carrier that changes frequency in a defined pattern. The 1997 IEEE 802.11 standard specifies a series of evenly spaced hop channels across the 2.4 GHz frequency band. These channels occur at 1 MHz intervals within the 2.4 GHz band.

The size of the 2.4 GHz frequency band and the number of hop channels are dependent on the regulatory agency rules in each geographic region. This is shown in Table 6-1. The IEEE 802.11 standard also specifies three hopping sequence sets for each region. These are shown in Table 6-2. The actual hopping patterns are derived from

Table 6-1

IEEE 802.11 frequency-hopping channels

Frequency	Frequency-Hopping Channels
North America 2.4–2.4835 GHz	79
Europe 2.4–2.4835 GHz	79
Spain 2.445–2.475 GHz	27
France 2.4465–2.4835 GHz	35
Japan 2.471–2.497 GHz	23

Table 6-2

IEEE 802.11 frequency-hopping sequences

Frequency	Channel Hop Sequences
North America 2.4–2.4835 GHz	Sequence 1: 0, 3, 6, 9, 12, 15, 18, 21, 24, 27, 30, 33, 36, 39, 42, 45, 48, 51, 54, 57, 60, 63, 66, 69, 72, 75 Sequence 2: 1, 4, 7, 10, 13, 16, 19, 22, 25, 28, 31, 34, 37, 40, 43, 46, 49, 52, 55, 58, 61, 64, 67, 70, 73, 76
Europe 2.4–2.4835 GHz	Sequence 3: 2, 5, 8, 11, 14, 17, 20, 23, 26, 29, 32, 35, 38, 41, 44, 47, 50, 53, 56, 59, 62, 65, 68, 72, 74, 77
Spain 2.445–2.475 GHz	Sequence 1: 0, 3, 6, 9, 12, 15, 18, 21, 24 Sequence 2: 1, 4, 7, 10, 13, 16, 19, 22, 25 Sequence 3: 2, 5, 8, 11, 14, 17, 20, 23, 26
France 2.4465–2.4835 GHz	Sequence 1: 0, 3, 6, 9, 12, 15, 18, 21, 24, 27, 30 Sequence 2: 1, 4, 7, 10, 13, 16, 19, 22, 25, 28, 31 Sequence 3: 2, 5, 8, 11, 14, 17, 20, 23, 26, 29, 32
Japan 2.471–2.497 GHz	Sequence 1: 6, 9, 12, 15 Sequence 2: 7, 10, 13, 16 Sequence 3: 8, 11, 14, 17

these sets. North America and Europe have 26 patterns within each sequence set, Spain has 9, France has 11, and Japan has 4.

Local authorities also regulate the hopping rate. In North America, the hopping rate is set at 2.5 hops per second with each transmission occupying a channel for less than 400 milliseconds. Channel occupancy is also called *dwell time*. In 2001, the *Federal Communications Commission* (FCC) proposed to amend its Part 15 rules to remove the minimum hop channel requirement of 75 channels and allow adaptive hopping techniques to be used. This rulemaking is designed to reduce interference with other systems operating in the 2.4 GHz frequency (such as Bluetooth). Studies have shown that up to 13 IEEE 802.11 FHSS systems can be colocated before frequency channel collisions become an issue.

The 1997 IEEE 802.11 standard defines two variations of GFSK as the modulation technique. Transmission at 1 Mbps requires the use of two-level GFSK. Transmission at 2 Mbps is obtained by optionally using four-level GFSK. Two-level GFSK modulation represents binary (0 or 1) values by deviating up or down from the base carrier frequency of the hop channel. The upper deviation represents a 1. The lower deviation represents a 0. Four-level GFSK is similar, but uses pairs of binary values (00,01,10,11) represented by two upper and two lower deviations within the hop channel. IEEE 802.11 GFSK uses transmission power inefficiently compared to the modulation techniques in IEEE 802.11 DSSS. This is one of the factors that limits the use of FHSS for higher-speed applications. The procedures used to produce a frequency-hopping signal are shown in Figure 6-2.

DHSS

DHSS systems mix high-speed bit patterns with the information being sent to spread the *radio frequency* (RF) carrier. Each bit of information has a redundant bit pattern associated with it, effectively spreading the signal over a wider bandwidth. These bit patterns vary in length and the rate at which they are mixed into the RF carrier. They are called *chips* or *chipping codes* and may vary in

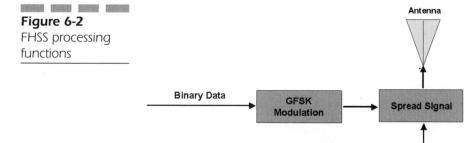

Figure 6-2
FHSS processing
functions

Frequency Hopping Spread Spectrum Transmission

length from as small as 11 bits to extremely long sequences. The speed at which they are transmitted is called the *chipping rate*. To an observer, these sequences appear to be noise and are also called *pseudorandom noise codes* (PNcodes). PNcodes are usually introduced into the signal through the use of hardware-based shift registers, and the techniques used to introduce them are divided into several groups including Barker codes, Gold codes, M-sequences, and Kasami codes. IEEE 802.11 uses a relatively simple 11-bit Barker code to spread the signal. IEEE 802.11b uses a group of Walsh-Hadamard codes also called *complementary codes*. DHSS systems such as cellular *Code Division Multiple Access* (CDMA) (IS-95) and the *Global Positioning System* (GPS) use Walsh and M-sequences codes.

These spreading codes also allow the use of statistical recovery methods to repair damaged transmissions. Another side effect of spreading the signal is lower spectral density—that is, the same amount of signal power is distributed over more bandwidth. The effect of a less spectrally dense signal is that it is less likely to interfere with spectrally dense narrowband signals. Narrowband signals are also less likely to interfere with a DSSS signal because the narrowband signal is spread as part of the correlation function at the receiver.

The frequency channel in IEEE 802.11 DSSS is 22 MHz wide. This means that it supports three nonoverlapping channels for operation. This is why only three IEEE 802.11 DSSS systems can be colocated. Figure 6-3 shows an IEEE 802.11 DSSS signal.

Figure 6-3
DHSS signal

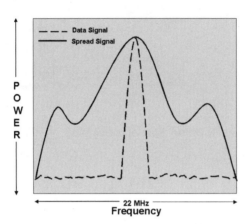

In addition to spreading the signal, modulation techniques are used to encode the data signal through predictable variations of the radio signal. IEEE 802.11 specifies two types of *differential phase-shift keying* (DPSK) modulation for DSSS systems. The first is BPSK and the second is QPSK. *Phase-shift keying* (PSK), as the name implies, detects the phase of the radio signal. BPSK detects 180-degree inversions of the signal, representing a binary 0 or 1. This method has an effective data rate of 1 Mbps. QPSK detects 90-degree phase shifts. This doubles the data rate to 2 Mbps. IEEE 802.11b adds CCK and *packet binary convolutional coding* (PBCC), which provide data rates up to 11 Mbps. Figure 6-4 shows the processing functions for IEEE 802.11 DSSS.

Many different modulation techniques were considered for the IEEE 802.11b specification. A compromise between Lucent and Harris Semiconductor resulted in the adoption of CCK as the 11 Mbps modulation technique. An advantage of CCK over some of the other proposals is its resistance to multipath distortion. PBCC was proposed by Texas Instruments and added as an optional modulation technique.

Figure 6-4
DHSS processing
functions

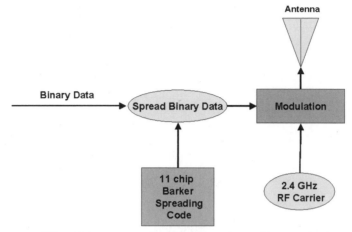

Direct Sequence Spread Spectrum Transmission

Orthogonal Frequency Division Multiplexing (OFDM) and IEEE 802.11a

IEEE 802.11a (5 GHz) uses OFDM as its frequency management technique and adds several versions of *quadrature amplitude modulation* (QAM) in support of data rates up to 54 Mbps. Bell Labs patented OFDM in 1970 and it is based on a mathematical process called *Fast Fourier Transform* (FFT). FFT enables 52 channels to overlap without losing their individuality or orthoganality. Overlapping channels is a more efficient use of the spectrum and enables them to be processed at the receiver more efficiently. IEEE 802.11a OFDM divides the carrier frequency into 52 low-speed subcarriers. Forty-eight of these carriers are used for data and four are used as pilot carriers. The pilot subcarriers allow frequency alignment at the receiver. Figure 6-5 shows an 802.11a OFDM signal.

One of the biggest advantages of OFDM is its resistance to multipath interference and delay spread. Multipath is caused when radio waves reflect and pass through objects in the environment. Radio waves are attenuated or weakened in a wide range depending on the object's materials. Some materials (such as metal) are opaque to

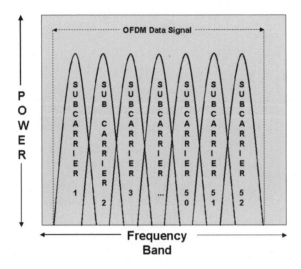

radio transmissions. As you can see, a cluttered retail store environment would be very different from an open warehouse environment for radio wave transmission and reception. This environmental variability is why it is so hard to estimate the range and data rate of an IEEE 802.11 system. Because of reflections and attenuation, a single transmission can be at different signal strengths and from different directions depending on the types of materials it encounters. This is multipath. IEEE 802.11a supports data rates from 6 to 54 Mbps. It utilizes BPSK, QPSK, and QAM to achieve the various data rates. This is shown in Table 6-3.

Table 6-3

IEEE 802.11
frequency-
hopping
sequences

IEEE 802.11a Data Rate	Modulation	Subcarrier Data Bits
6 Mbps	BPSK	1
9 Mbps	BPSK	1
12 Mbps	QPSK	2
18 Mbps	QPSK	2
24 Mbps	QAM-16	4
36 Mbps	QAM-16	4
48 Mbps	QAM-64	6
54 Mbps	QAM-64	6

Delay spread is associated with multipath. Because the signal is traveling over different paths to the receiver, the signal arrives at different times. This is delay spread. As the transmission rate increases, the likelihood of interference from previously transmitted signals increases. Multipath and delay spread are not much of an issue at data rates less than 3 or 4 Mbps, but some sort of mechanism is required as rates increase to mitigate the effect of multipath and delay spread. In IEEE 802.11b, it is CCK modulation. In IEEE 802.11a, it is OFDM. The IEEE 802.11g specification also uses OFDM as its frequency management mechanism.

The adoption and refinement of advanced semiconductor materials and radio transmission technologies for IEEE 802.11 provides a solid basis for the implementation of higher-level functions. The next step up the protocol ladder is the definition of access functionality. Without structured access, the physical medium would be unusable. Medium access is discussed in Chapter 7.

Medium Access Control (MAC) Concepts and Architecture

The IEEE 802.11 *Medium Access Control* (MAC) is common to all IEEE 802.11 *physical* (PHY) layers and specifies the functions and protocols required for control and access. The MAC layer is responsible for managing data transfer from higher-level functions to the physical media.

MAC Layer Services

Devices using the IEEE 802.11 PHY and MAC as part of a *wireless LAN* (WLAN) are called *stations*. Stations can be endpoints or *access points* (APs). APs are stations that act as part of the *distribution system* (DS) and facilitate the distribution of data between endpoints. The MAC provides nine logical services: authentication, deauthentication, association, disassociation, reassociation, distribution, integration, privacy, and data delivery. An AP uses all nine services. An endpoint uses authentication, deauthentication, privacy, and data delivery. Each service utilizes a set of messages with information elements that are pertinent to the service. These services are defined in Table 7-1.

MAC Layer Architecture

Both the PHY and MAC layers are conceptually divided into management and data transfer capabilities. The PHY management capability is provided by the *PHY layer management entity* (PLME). The MAC management capability is provided by the *MAC layer management entity* (MLME). The PLME and MLME exchange information about PHY medium capabilities through a *Management Information Base* (MIB). This is a database of physical characteristics such as possible transmission rates, power levels, antenna types, and so on. Some of these characteristics are static and some can be changed by a management entity. These management functions support the main purpose of the MAC, which is to transfer data elements. These data elements originate in the *Logical Link Control* (LLC) layer. Packages of data passed to the MAC from the LLC are called *MAC service data units* (MSDUs). In order to transfer the MSDUs to the PHY, the MAC uses messages (frames) containing functionally related fields. There

Table 7-1

IEEE 802.11 MAC
services

MAC Service	Definition	Station Type
Authentication	Establishes the identity of a station and authorizes it to associate.	Endpoint and AP
Deauthentication	Removes an existing authentication.	Endpoint and AP
Association	Maps a station to an AP and enables the AP to distribute data to and from the station.	AP
Disassociation	Breaks an existing association relationship.	AP
Reassociation	Transfers an association between APs.	AP
Privacy	Prevents unauthorized viewing of data through the use of the *Wired Equivalency Protocol* (WEP) algorithm.	Endpoint and AP
Distribution	Provides data transfer between stations through the DS.	AP
Data delivery	Provides transfer of data between stations.	Endpoint and AP
Integration	Provides data transfer between the DS of an IEEE 802.11 LAN and a non-IEEE-802.11 LAN. The station providing this function is called a *portal*.	AP

are three types of MAC frames: control, management, and data. One of these messages is called a *MAC protocol data unit* (MPDU). The MAC passes MPDUs to the PHY layer through the *Physical Layer Convergence Protocol* (PLCP). The PLCP is responsible for translating MPDUs into a format that is *physical medium dependent* (PMD). The PMD layer transfers the data onto the medium. For example, you have very different PMD layers for infrared and 2.4 GHz radio. The flow is reversed for incoming data. Figure 7-1 shows the relationship between the LLC, MAC, and PHY layers.

MAC data transfer is controlled through two distinct coordination functions. The first is the *distributed coordination function* (DCF), which defines how users contend for the medium as peers. DCF data transfers are not time sensitive and delivery is asynchronous. The second is the *point coordination function* (PCF), which provides centralized traffic management for data transfers that are sensitive to delay and require contention-free access.

Figure 7-1
MAC/PHY
interses

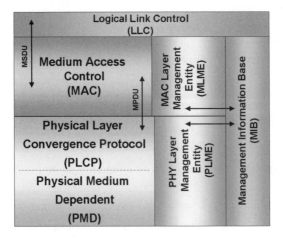

Figure 7-1
MAC/PHY
interfaces

Distributed Coordination Function (DCF)

The DCF defines how the medium is shared among members of the wireless network. It provides mechanisms for negotiating access to the wireless medium as well as mechanisms for reliable data delivery. One of the fundamental differences between wired and wireless media is that it is difficult to detect and manage data collisions on wireless media. The primary reason for this is that stations in a radio network are not guaranteed to hear every other station's transmissions. This is typically the case when an AP is used in IEEE 802.11's infrastructure *basic service set* (BSS) and is called the *hidden-node problem*. In Figure 7-2, Node A and Node B can hear each other and the AP. Node B and Node C can hear each other and the AP. However, Node A and Node C cannot hear each other. If they attempt to use the medium simultaneously, a collision will corrupt the data transmission. To address this problem, the DCF has implemented a collision avoidance function.

Collision Avoidance Mechanisms

In order to avoid collisions, the DCF uses mechanisms for sensing whether the medium is in use before transmitting. If the medium is in use, the station will wait according to a predetermined algorithm before attempting to transmit. This method of sensing the medium

Figure 7-2
Hidden network
nodes

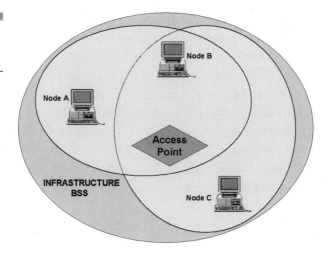

and waiting before transmission is called *carrier sense multiple access collision avoidance* (CSMA/CA). The DCF supports complementary physical and virtual carrier sense mechanisms.

Because each medium has different characteristics, physical sensing of the medium is a function of the PHY being used. Physical sensing of the medium is called *clear channel assessment* (CCA). For example, a direct sequence radio PHY can be directed to report the medium to be in use in any of three separate conditions. The first condition reports an in-use condition if any energy above a defined threshold is detected on the medium. The second condition reports an in-use condition if any *direct sequence spread spectrum* (DSSS) signal is detected. The last condition reports an in-use condition if a DSSS signal above a defined threshold is detected on the medium. Physical sensing is very efficient, but is susceptible to the hidden-node problem (you can't sense something that is out of range).

In virtual carrier sensing, no actual physical sensing of the medium occurs. Instead, information about the use of the medium is exchanged through the use of control frames. As opposed to physical carrier sensing, virtual sensing greatly reduces the probability of collisions between hidden nodes. It also reduces the overall throughput. This is due to the additional control frames that must be exchanged. Because this overhead is fixed, the smaller the data frames being sent, the higher the percentage of overhead that is added. In networks with a large amount of small packets or low collision rates, it may be best to only use physical sensing. For this reason, the DCF's virtual carrier sensing mechanism is optional. The virtual carrier sense control messages are called *Request to Send* (RTS) and *Clear to Send* (CTS) frames. A frame size threshold (RTS threshold) can be set that enables a virtual carrier sense procedure only for packets greater than a specified size. The RTS/CTS procedure is not used for broadcast or multicast frames (single frame with multiple destinations) because this could generate multiple conflicting CTS responses. The virtual carrier sense mechanism also helps to avoid collisions when two overlapping BSSs utilize the same radio channel for transmission.

In Figure 7-3, when Node A wants to send data, it sends an RTS frame to the AP with addressing and timing information. It sends the address of the node that will receive the impending data frame (*receiver address* [RA]), its own address (*transmitter address* [TA]), and how long it wants to transmit (duration). The calculation of the duration has several elements. An AP receiving an RTS frame replies with a CTS frame. The CTS can be heard by all nodes within the AP's range. In forming the CTS frame, the AP copies the TA from the RTS into the RA of CTS frame. It also copies the duration field into the CTS after adjusting it for the actual transmission of the CTS. The receipt of a CTS causes the receiver to store the duration field as its *network allocation vector* (NAV). The NAV is a timer that

Figure 7-3
Virtual carrier sense

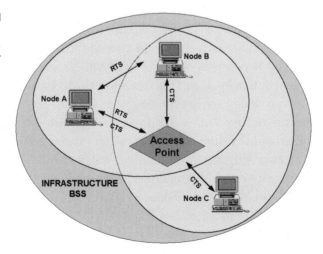

indicates the amount of time that remains before the medium can be used. This value counts down on a regular basis, and when it reaches zero, it indicates that the medium is free. It is updated every time an RTS or CTS with a larger value is received. By combining the physical sensing of the medium with the RTS/CTS procedure, it is possible for a hidden node that is unable to receive from the originating node to avoid collisions with an impending data transmission.

In addition to the RTS and CTS control frames, the DCF CSMA/CA procedure requires an *acknowledgment* (ACK) frame to be sent upon successful receipt of certain types of frames. There is no *negative acknowledgement* (NACK), only a timer that indicates how long to wait for an ACK before the transmission is assumed to be in error. The DCF also provides several frame interval timers based on PHY-specific values. These interval timers represent the time that a station must sense that the medium is idle before starting a transmission. There are two PHY-specific intervals that serve as the basis for the other frame interval timers: the *slot time* and *short interframe space* (SIFS). The slot time for a DSSS PHY (20 microseconds) is defined as the sum of the receive-transmit turnaround time and the energy-detect time including any propagation delay. The slot time for the IEEE 802.11 frequency-hopping PHY is 50 microseconds. The SIFS is the shortest of the frame interval spaces and is used to allow the completion of an in-progress transmission. The SIFS for the DSSS PHY is 10 microseconds. The SIFS for the FHSS PHY is 28 microseconds.

The slot time and the SIFS are used as components in three other frame intervals. These are the *distribution interframe space* (DIFS), the *extended interframe space* (EIFS), and the *PCF interframe space* (PIFS). The DIFS is used by the DCF to enable the transmission of data and management MPDUs. The EIFS is used to enable the processing of frames reported to be erroneous by the PHY layer. The PIFS enables a station to have priority access to the medium when operating in the PCF contention-free mode.

One other timer is used in the DCF virtual CSMA/CA capability: the *backoff interval*. If a station wanting to transmit detects that a transmission is in progress, it will wait before retrying the transmission. The time it will wait is determined by the *backoff algorithm*, which is an exponential progression between minimum and maximum values. The starting value for this progression is calculated by multiplying a random number between the minimum and maximum backoff values with the slot time of the PHY. The backoff time is subsequently calculated as sequentially ascending integer powers of 2, minus 1. For example, if the random value is 3 and the slot time is 10 microseconds, the station would wait 7 or $(2^3 - 1) \times$ 10 microseconds (70 microseconds). The retries would then continue using 15 $(2^4 - 1)$, and then 31 $(2^5 - 1)$ up to the maximum value between retries. Because a random number is used, two stations entering a transmission retry sequence will usually not arrive at the same backoff interval. This prevents two stations from repeatedly colliding because their retry sequences become synchronized. The station also has a retry counter that can limit the number of retries. Figure 7-4 illustrates the virtual carrier sense protocol.

Figure 7-4

Virtual carrier sense protocol

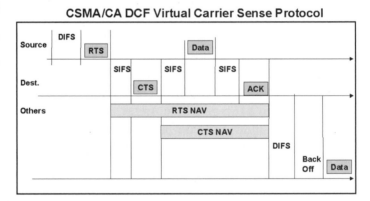
CSMA/CA DCF Virtual Carrier Sense Protocol

The DCF Carrier Sense Protocol is a robust method of overcoming the challenges of radio data transmission between network peers. Centralized traffic management such as that provided by an access point is discussed next.

Point Coordination Function (PCF)

The PCF polls associated stations and manages frame transmissions on their behalf. A station performing PCF traffic management is called a *point coordinator* (PC). The PCF is an optional capability that provides connection-oriented services for delay-sensitive traffic. The PCF is more complex to implement, but it provides a moderate level of priority frame delivery for time-sensitive transmissions. The PCF is one of the basic methods utilized in the work of Task Group *e* (IEEE 802.11e).

The PC uses beacon signals to broadcast a duration for a contention-free period to all associated stations. This causes them to update their NAV and wait for the duration of the contention-free period. In addition, stations must wait the PIFS interval to further decrease the possibility of data collisions. The transmission of the additional polling and ACK messages required by the PCF is optimized through piggybacking multiple messages in a single transmission. For example, the PC may append both ACKs of previous transmissions and polling messages for new traffic to a data frame. This enables the transmission to avoid waiting the interframe interval specified for individual frame transmissions.

Power Management and Time Synchronization

In addition to the CSMA/CA control frames (RTS, CTS, ACK, and contention polling), the MAC also provides control frames for power management and time synchronization. APs provide a time synchronization beacon to associated stations in an infrastructure BSS. In an independent BSS where stations are operating as peers, an algorithm is defined that enables each station to reset its time when it receives a synchronization value greater than its current value. Stations entering a power-save mode may inform a PC through the frame control field of a message. The AP will then buffer transmissions to the station. A station is informed that it has buffered transmissions waiting when it wakes periodically to receive beacon frames. It can then request transmission. A station in active mode can receive frames at any time during a contention-free period. A station in a power-save mode will periodically enter the active mode to receive beacons, broadcast, multicast, and buffered data frames.

Data Fragmentation

The longer a transmission lasts, the greater the probability that it will be corrupted by interference. To allow the transmission of shorter frames and reduce the likelihood of interference, the IEEE 802.11 MAC provides a method of breaking transmissions into smaller units. This is called *fragmentation*. A value called the *fragmentation threshold* specifies that frames over a specified size should be divided into multiple transmissions. The frame header contains a sequence control field that shows the order of the fragments. Fragments constituting a frame are transmitted immediately after one another without any contention for the medium. Each fragment has its own *cyclic redundancy code* (CRC), and an individual ACK is transmitted for each fragment. The fragment transmissions are separated by the appropriate frame interval spaces. The transmission of a sequence of fragments is called a *frame burst*. If an error occurs on a fragment, subsequent fragments are not transmitted until the previous frame is acknowledged. The retransmission and backoff rules apply to fragmented frame transmissions. Duration information in the fragments and ACK frames sets the NAV. Broadcast and multicast frames are not fragmented even if their size exceeds the fragmentation threshold. Fragmentation is shown in Figure 7-5.

Figure 7-5
Frame
fragmentation

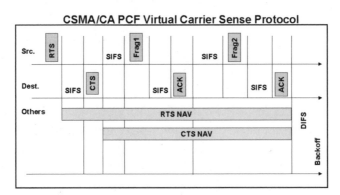

Through the use of CSMA/CA and the definition of rules for peer-to-peer as well as centrally managed data transfer, the MAC layer provides reliable structured access to the PHY layer. Physical, diosynchrosies are masked from the upper layers enabling the logical link control functions and the whole suite of TCP/IP protocols.

Wireless LAN (WLAN) Security

Much has been written about *wireless LAN* (WLAN) security. Most works detail the shortcomings of the *Wired Equivalency Protocol* (WEP) algorithm. This chapter will offer a broader view of security. We will explore the current state of the 802.11 standards. We will also discuss the security policies and, more specifically, the security measures and mechanisms in a wireless environment. We will also examine the expected future enhancements to the 802.11 standard.

Architecture and Terminology

Although there are differences in the technology, hardware, internal algorithms, and vendor implementations, security principles remain the same whether you are discussing mainframes, *personal computers* (PCs), wired LANs, or WLANs. There have been reported hacks and security breaches in virtually every technology introduced to date. Although it is true that the wireless medium has some inherent risks and that the encryption algorithm needs to be enhanced, saying that a given technology is inherently unsafe is misleading. By taking care to analyze the needs of the organization to protect its assets and implementing the proper security mechanisms correctly, you will do more to ensure a successful outcome than the use of any particular technology.

The most widely cited security risk with wireless networks is that because they radiate data over the airwaves in a pattern that may be beyond the physical boundaries that the company can control, these radio waves and the data they carry can be passively intercepted. Although this is true, there are installation techniques that can limit the area over which the radio waves radiate. This problem also exists in wire networks, but the wired network is presumed to be secure because it is within the confines of the building perimeter. The electromagnetic waves emitted by the wired network can be intercepted by using sensitive listening equipment, although the person would have to be fairly close to the network.

Most LAN adapters for both wired and wireless networks offer a promiscuous mode that will enable you to capture every packet on that segment of the LAN. Furthermore, data on the wired LAN is often presumed to be secure because in order to access the network, a person needs to be physically wired into the network. This view fails to take into account that if corporate users can access the Internet, then hackers from outside can get in if proper precautions are not taken. This view also does not take into account the issues raised when corporate users access the corporate network while traveling or telecommuting.

To understand the security implications of implementing a WLAN, we need to first discuss some basic architecture and terminology. Recall that in Chapter 2, "The IEEE 802.11 Alphabet Spelled Out," we introduced the concepts of an *independent basic service set* (IBSS), an infrastructure *basic service set* (BSS), and an *extended service set* (ESS). When two or more *stations* come together and communicate directly, it is referred to as an IBSS. An IBSS is illustrated in Figure 2-1. A BSS is formed when a device called an *access point* (AP) is added to provide centralized communication distribution, buffering, and gateway functionality. Because the station can be turned off or on and move either in or out of range, the association between the stations and BSSs is dynamic. A BSS is illustrated in Figure 2-2. In order to provide sufficient coverage, it may be necessary for two or more BSSs to form an extended network. The component used to interconnect the BSSs is called the *distribution system* (DS). Access to the DS is provided by an AP, which is an addressable STA that provides DS services in addition to acting as a STA. Connecting multiple BSSs through the DS creates an extended network of arbitrary size and an ESS is formed. An ESS is illustrated in Figure 2-3. Lastly, for a BSS or ESS to communicate with the corporate wired network, a *portal* is added, which is the logical connection point to the wired network.

By comparing a wireless network to a wired network, we can draw the following comparisons: In a traditional wired LAN, an address is equivalent to a fixed physical location. In a WLAN, a STA is an addressable unit, which is a message destination, but is not necessarily a fixed location. Wired LANs are made up of static, but movable components, whereas WLANs have dynamic topologies (recall the discussion on mobility versus portability in Chapter 2). Wired

LANs have fixed physical boundaries, such as walls. WLANs radiate signals outward and upward using time-varying, asymmetric, and three-dimensional propagation properties and are not constrained by walls, ceilings, or other physical boundaries. Furthermore, WLANs are unprotected from outside signals. Wired LANs have full connectivity with each device on the network. In a WLAN environment, any given station may or may not be able to hear any other station at any particular time. The three basic functions of security are authentication, integrity, and confidentiality. *Authentication* is used to determine that network access is granted only to approved persons (or devices). *Integrity* is the means by which the recipient of the data transfer can know that the data has not been tampered with. *Confidentiality* is the means by which data being transmitted is kept secret until it reaches its destination. In other words, we want to know with whom we are communicating, that our messages arrive intact, and that no one else can hear or read what we have said. As we can see from the inherent physical property differences used in wireless networks, ensuring that "air" is secure can be quite a challenge. Let's take a look at the wireless standard and discuss the implications of using wireless technology within several broad categories of security risk.

The 802.11 Standard

Given the inherent differences in the PHY layer and the fact that IEEE 802.11 is required to appear to be a current-style IEEE 802 LAN to the higher layers, the standard defines untraditional functionality in the MAC layer. The 802.11 standard provides for three basic security mechanisms. The first is the *Service Set Identifier* (SSID), which can be used as a simple password between the STA and AP or can be used as a broadcast location identifier in a public network. The second is a list of MAC addresses that can be manually entered and managed to limit access to only the addresses on the list. The last is the WEP encryption algorithm. The next three sections discuss each of these security mechanisms in further detail.

Service Set ID (SSID)

The SSID is an alphanumeric code that is entered into each AP and STA that participates on the same WLAN. The AP will, by default, broadcast the SSID periodically. Many APs allow the administrator to disable the broadcast of the SSID, thereby forcing users to either know the network name or to have software to capture this information. As we will see in a minute, this is not a strong form of security because the SSID is sent in cleartext as a reply to a probe from a station even if it is not broadcast. In fact, Windows XP ships with client software that is capable of detecting this value and will list all networks in the area for the user to choose from.

MAC Address Lists

The second mechanism provided is the MAC address list. This list contains the MAC addresses of the wireless *network interface cards* (NICs) that may associate with any given AP. Some vendors provide tools to automate the entry and update processes; otherwise, this is an entirely manual process. This is also not a very strong security mechanism. Because MAC addresses are also transmitted in cleartext, it is easy to discover MAC addresses with a sniffer. As one of the attacks discussed in the following section describes, a wireless NIC can be configured with a known good MAC address that has been sniffed, and an attacker can gain access to the network.

The WEP Algorithm

Once a computer has been authenticated and granted access to the network, data will presumably be transmitted to or from the station. Because the data is being transmitted via radio waves, it can be easily intercepted, making it important for the data to be encrypted. The IEEE 802.11 standard provides WEP as an encryption algorithm. The WEP algorithm uses the RC4 *pseudorandom number generation* (PRNG) algorithm originally developed in 1987 and licensed by RSA Data Security, Inc.

The WEP algorithm takes a block of plaintext and bitwise XORs it with a pseudorandom key sequence generated by the algorithm to produce the ciphertext message to be transmitted. The pseudorandom key sequence generation begins by using a secret key and an *initialization vector* (IV). A secret key, which has been distributed to cooperating STAs, is concatenated with an IV to produce a seed. The seed is then used as input to the PRNG. The PRNG then outputs a key sequence equal in length to the number of data octets that will be transmitted, plus four octets that will be used to protect the *integrity check value* (ICV). WEP uses the CRC-32 algorithm to operate on the plaintext block to generate a 32-bit ICV. The key sequence generated from the PRNG is then XORed with the plaintext block concatenated with the ICV to produce the ciphertext. The act of concatenating the IV, followed by the *MAC protocol data unit* (MPDU), followed by the ICV generates the frame body. This process is shown in Figure 8-1.

Once a message arrives, the IV of the incoming message is combined with the secret key, which has been distributed to all cooperating STAs, to generate the seed that will be used in the PRNG operation. The WEP PRNG operation is then performed to produce the key sequence needed for decryption. Because the algorithm uses a symmetric scheme where the same key is used for the encryption and decryption of data, the ciphertext received in the message is XORed to yield the original plaintext and ICV. The CRC-32 algo-

Figure 8-1
WEP encryption

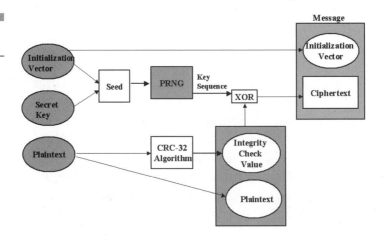

rithm can then be used on the recovered plaintext and compared to the ICV transmitted with the message. The decryption process is shown in Figure 8-2.

Several things should be noted about this process as specified in the IEEE 802.11 standard. First, WEP is optional. The delivered default is for WEP to be disabled. However, if encryption is desired, the WEP algorithm is the only Wi-Fi–supported algorithm. Second, although we have discussed a single secret key, the standard permits up to four secret keys to be defined. Third, the 802.11 standard relies on external key management services to distribute the secret keys to each station; it does not, however, specify how these key distribution services should be implemented. Fourth, the secret keys are under the control of the network administrator and remain static unless changed by the network administrator. Obvious problems result from the static nature of the keys and the manual process of key management as changing the keys on each station in a large network can be extremely time consuming. If a station is lost due to theft or accident, the keys will need to be changed on all stations; however, given the mobility of the population and without a convenient way to manage this task, the network administrator may be under great pressure to accomplish this in a reasonable timeframe. Fifth, the IV changes periodically; however, the interval at which it changes and the manner in which it is incremented is left up to the implementers. The IV is transmitted in the clear because it must be known by the recipient in order for the received message to be successfully decoded.

Figure 8-2
WEP decryption

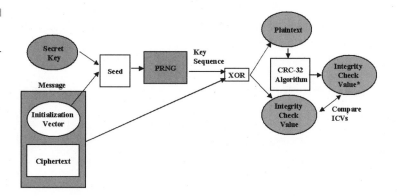

Putting It in Place—Services and Messages

Recall that the three basic security functions are authentication, integrity, and confidentiality. The 802.11 standard includes specifications at the MAC and PHY levels to implement security through the use of authentication, association, and privacy services. The IEEE 802.11 standard specifies two categories of services in the MAC layer to accomplish the functionality of the DS; however, it does not explicitly specify the details of DS implementations. Every 802.11-conformant station must provide the station services that are shown in Table 8-1. DS services, which are also shown in Table 8-1, are provided by the DS and may or may not be within a physical AP. Of the nine services specified by IEEE 802.11, three are used for controlling LAN access and confidentiality, and the other six are used for message delivery between stations. Each service is supported by either management or data messages and by one or more MAC frame types. Frame types and message formats were discussed in detail in Chapter 7, "Medium Access Control (MAC) Concepts and Architecture." The services involved in controlling LAN access and confidentiality are shown in Table 8-1.

The process of connecting a wireless client to a network begins with the client broadcasting probes containing the client MAC address and the *BSS identifier* (BSSID) on all *radio frequency* (RF) channels used by 802.11b. All APs within range will respond with their own BSSID, MAC address, and channel. The client can then limit its signal to the proper channel and begin the process of

Table 8-1

Station and DS services

Station Services	DS Services
Authentication	Association
Deauthentication	Disassociation
Privacy	Distribution
MAC service data unit (MSDU) delivery	Integration
	Reassociation

authentication. The 802.11 standard provides two methods of authentication. In open system authentication, which may violate implicit assumptions made by higher network layers, any client can authenticate as long as the station receiving the request to authenticate does not decline and the station passes any MAC address filtering rules that have been set up. Because WEP is delivered disabled by default, open system authentication is the default method and all authentication packets are transmitted without encryption.

In the second method, shared key authentication, WEP is required to be enabled, and identical WEP keys are required to have been previously installed on the client and AP. The initiating station requests shared key authentication. The AP returns 128 bytes of randomly generated unencrypted challenge text. The initiating station then encrypts the text using the shared key and returns the WEP-encrypted data. The AP verifies the validity and integrity of the data, and then authenticates the connection and confirms the authentication with the client. Successful completion of the process ensures that both stations' secret keys match. The authentication provided in the standard is link level. The standard does not provide end-to-end or user-to-user level authentication, but it recommends that IEEE standard 802.10 be implemented if this level of authentication is desired. After authentication is complete, the client initiates the association process.

In order for a message to be delivered, the distribution service must know which AP to access for each station. A station can only be associated with one AP at any given time. A mutually acceptable level of authentication must be established before association with an AP can be established and DS services can be obtained. The client transmits its BSSID, which is verified by the AP. With a positive match, the AP adds the client to its table of authenticated clients and returns an affirmation to the client. At this point, the client is now connected to the network. This process is shown in Figure 8-3. Given the mobility of clients within the WLAN, a reassociation service is also provided to facilitate the move from one AP to another within the ESS. Association and reassociation are DS services.

The deauthentication service is a station service just like the authentication service and is used to terminate an existing authentication. Deauthentication will also cause a station to be disassociated

Figure 8-3
Station
authentication/
association

Request for shared key authentication
Unencrypted challenge text
WEP-encrypted data
Authentication Response

Association Request – ESSID
Association Response

Figure 8-4
Station
deauthentication/
disassociation

Disassociation Notification

because association is a requirement for authentication. Deauthentication is a notification—not a request—and may be initiated by either party (STA or AP) and cannot be refused by either party. This process is shown in Figure 8-4.

The last service is the privacy service, which is also a station service. The privacy service is used to invoke the use of the WEP encryption algorithm and encrypt the MPDU. Through the use of certain *Management Information Base* (MIB) attributes, encryption is turned on and the correct shared key is used to encrypt the data prior to transmission.

Much has been written about wireless security in general and more specifically about WEP and its shortcomings. The next few sections will discuss these widely publicized shortcomings, what can be done to compensate for them, and the anticipated future enhancements to the standard.

Security Risks

For both wired LANs and WLANs (as well as most other technologies), several categories of security issues violate the three basic functions of security—authentication, integrity, and confidentiality. The four broad categories of risk that a network manager should be concerned with are the loss of equipment, a virus infection, misuse by authorized personnel, and attacks by unauthorized personnel. Depending on the technology, each of these broad categories can be more or less damaging and more or less likely.

Loss of Equipment

The loss of equipment is an issue that has recently received quite a bit of attention due to events within the FBI. The loss of a laptop or other piece of equipment poses the issue of what data is contained within the device. It is possible for an unscrupulous person to dial into the wired network using lost or stolen equipment and stored passwords, and masquerade as an authorized user. This scenario is possible with current wired networks and is not dependent upon having access to a WLAN.

The loss of a device equipped with wireless access certainly carries the same risks. Many would argue that the loss of a wireless-equipped device could make it easier to access the LAN because it could be accomplished from the parking lot, but access to a computer with stored authentication information to a wired LAN makes access possible from around the globe. The difference is that intercepting traffic using a wireless device from the parking lot is virtually undetectable and assuming that the company has a robust monitoring system in place, dialing into the wired network would be detected. The question is whether it would be too late.

Virus Infection

Virus infection is another issue that affects both wired and wireless networks. To date, there have been no reported viruses that infect cell phones; however, there have been viruses that are capable of sending text messages to cell phones. Two of these are VBS/Timo-A and the LoveBug. There have been reports of viruses that infect Palm OS as well as viruses carried on diskette, CD-ROM, and e-mail. These viruses can infect laptops whether or not they are wireless equipped and can be introduced into and spread via either the larger wired or wireless network. Although the reported Palm

OS viruses appear to pose little threat at this point, it makes sense to protect all of your mobile devices and, by extension, your entire corporate network.

Misuse by Authorized Personnel

Misuse by an authorized person, whether intentional or accidental, is by far the most difficult threat to protect against. Once access is granted to the system, what that user does with his or her access is very difficult to control. Disgruntled employees have been known to steal company data, delete files, change sensitive data, and perform any number of other malicious deeds. Even if an employee is not intending to do any harm, he or she may do so unintentionally by sharing passwords, miskeying data, printing on the wrong printer, sending e-mail containing sensitive data, downloading data to a hard disk or a floppy disk, and doing any number of other seemingly innocuous things. The best protection for this type of action is to have a set of security policies and monitor the users to ensure that the policies are consistently being followed. However, monitoring requires a good deal of time, discipline, and an understanding of the data, and authorized ownership and uses. Monitoring with its over-tones of Big Brother is usually not done (or at least not very thoroughly) because most people assume that if a user has been granted access to the system, he or she is not doing anything wrong. Monitoring is a task that should be delegated to or shared by the data owner or the corporate security officer and not left solely to the network administrator.

One example of this type of risk can be illustrated by the use of the German Enigma machine during World War II. The Germans were convinced that by using the Enigma machine, their codes were unbreakable. The machine looked like a typewriter, but contained a series of rotors connected by wiring. One rotor's electrical output was the input to the next rotor. As the operator pressed each key, one or more rotors moved and different electrical circuits connected through the unit. This resulted in the same letter being coded to a different substitute letter. For example, AAA might be translated as XUR instead of RRR. This made it much more difficult to break the code by using simple pattern recognition. The machine also was equipped with an adjustable ring on each rotor, which determined when its neighboring rotor would rotate. The Germans also added another layer of substitution by adding what was known as a *Stecker board*, which substituted predetermined keyboard letters with another letter prior to being sent to the rotors.

The combination of the rotor sequence, the ring settings, and the Stecker board was known as the *key*. A reflecting mechanism sent the electrical signal back in the opposite direction, making the machine reciprocal. This allowed encoding and decoding while using the same key settings. Because the Germans relied on the use of radio communications to keep in contact with the field commanders, the Enigma machine was instrumental in keeping the airwaves and the messages secure. The Germans had designed the machine to be secure, even if one or more fell into enemy hands. The keys were usually changed daily and the keys were sent to the units by courier each month. The sheer number of combinations of keys made the task of breaking the code virtually impossible. However, the Allies did crack the code, but this was mainly due to errors made by the people operating the machines.

The Polish Cypher Bureau purchased a commercial version (which did not have the Stecker boards) of an Enigma machine in the 1920s. The French Intelligence Service had obtained a booklet from a German traitor that described the Enigma setup procedures and gave it to the Poles. The German traitor also later provided old and seemingly useless messages in both plain and coded format along with the starting keys. One of the Poles, Marian Rejewski, was able to understand the wiring of the rotors used by setting up mathematical equations. Now that the Poles had a working model of the German Enigma, only the daily key was needed to decipher the messages. Thanks to a few procedural flaws and human errors, messages could be decoded.

One example was that even though the Enigma machine was set to a predetermined daily key, the operator was allowed to select a message setting, which was supposed to be random. Some operators, however, were using easy-to-guess settings such as keyboard shortcuts, repetitions, or their girlfriends' initials. Some operators could be identified by the message settings that they repeatedly used (known as their *fists*). By knowing the operator and his message settings, the Allies were sometimes able to group messages together and figure out the daily key. Also, by identifying the operator, many times the unit would be known. The Germans included a "to" and "from" line in their messages. Knowing parts of the plain message makes cracking the code much easier. One German operator would transmit "nothing to report" each day if no other message was to be sent using the daily key. Some careless operators upon learning that they had forgotten to reset the daily key would resend the identical message after resetting the key to its proper sequence, enabling the Allies to compare the identical message.

Several mistakes that apply to all security systems can be learned from this example. First, don't assume that there are no traitors out there who could do you harm. Second, the operators were not properly trained on the procedures and the importance of consistently following them. Third, transmitting redundant information gave the Poles valuable information and allowed them to ultimately crack the code. Lastly, even though German weather ships were captured along with the key sheets for the month, the German leaders would not believe that their code was being broken. The Germans assumed that the crew would destroy this sensitive data and did not change the keys after such an incident. The Enigma machine codes could have been unbreakable if not for the errors made largely by authorized personnel.

Attacks by Unauthorized Personnel

Attacks by unauthorized personnel can take many forms. These types of attacks are usually the most obvious and in many cases they are widespread in nature, such as virus attacks or bugs in a major hardware or software release. They are also generally reported in the mainstream media and/or by the vendor in the trade press. The most common types of attacks are described in the following paragraphs.

Eavesdropping   Eavesdropping is the passive acquisition of information from a network. Just as you can listen to other people's conversations, information can be overheard on the network. This method of gathering information about the network is getting easier with the release of several products. Airopeek, Airsnort, NetStumbler, and WEPCrack are all programs that enable you to acquire information such as the SSID, the MAC address of the AP, and information about whether WEP is enabled. These programs can also be used to intercept wireless network traffic by analyzing the packets to determine login IDs, passwords, or other sensitive corporate data.

Insertion Attacks   Configuring a device to gain access to a network or inserting unauthorized devices into a network in order to gain access is called an *insertion attack*. By installing wireless network cards and being in the vicinity of a target network, a device can be configured to gain access. Unauthorized APs can be installed in an attempt to get users to connect to a hacker's AP rather than to the intended network AP. If these APs are installed behind the corporate firewall, the risk of attack is much greater. This can sometimes be done by well meaning, but misinformed employees.

Man-in-the-Middle Attacks   In order to execute a man-in-the-middle attack, two hosts must be convinced that the computer in the middle is the other host. The classic version of this attack occurs when an attacker intercepts packets from the network, modifies them, and reinserts them back into the network.

Spoofing   *Spoofing* is pretending to be someone or something that you are not, such as using another person's user ID and password. *Domain Name Service* (DNS) spoofing is accomplished by sending a DNS response to a DNS server on the network. *Internet Protocol* (IP) address spoofing depends on the fact that most routers only look at the destination IP address, not the sending address. Validating the sending IP address can prevent this type of spoofing.

Brute-Force Password Attacks   Also known as *password cracking*, or *dictionary attacks*, this type of attack uses a dictionary and repeated attempts to test passwords to attempt to gain access to the network. This type of attack is possible even if password authentication is implemented.

Denial of Service Attacks   Denial of service attacks do not allow a hacker to gain access to the network; rather, they prevent the network from providing services to anyone. Usually this is accomplished by overloading a resource to cause a failure. The overload causes the host to become unavailable, much like those annoying messages of "all circuits are busy." There are many variations of these types of attacks depending on the type of resource blocked (disk space, bandwidth, internal memory, and buffers), and some are more easily prevented than others. In the simplest case, turning off the service when it is not needed prevents this type of attack. In other cases, they cannot be easily blocked without limiting the use of a necessary resource. In a wireless network, because the airwaves are shared by other devices such as cordless telephones, microwave ovens, and baby monitors, an attacker with the proper equipment can flood the airwaves with noise and disrupt service to the network.

Reported Flaws

There have been a number of widely publicized theoretical analyses as well as reports of successful attacks on WEP. We will outline several of them in this section; however, for specific details, please see the referenced works.

The *Internet Security, Applications, Authentication, and Cryptography* (ISAAC) team at the University of California at Berkeley published a study in January 2001 detailing four possible attacks against WEP. These attacks take advantage of two weaknesses in WEP. The first centers on the limitations of the IV—specifically on the fact that a network may cycle through all possible IVs in two to five hours. This, coupled with the use of static keys, results in a highly likely scenario of collisions. The second type of weakness centers on the CRC-32 checksum algorithm and its linear properties. Using this algorithm, it is possible to change data bits in the message and recompute an appropriate matching ICV.

The first attack is a passive attack that enables the decryption of network messages. Because the IV in WEP is a 24-bit field, all IVs within the space will be used within a relatively short period of time (two to five hours) and reuse of the IV will then be mandatory. Also, because the keys are static and are difficult to update in practice, IV collisions, where the same IV and key are sent in different frames, may occur fairly frequently on the network. If an attacker were to intercept wireless traffic, watching for IV collisions to occur, he or she could then XOR pairs of packets with the same IV to determine the exact contents of the messages. As an alternative option, the attacker could send known traffic from the outside to the wireless network. The attacker could then intercept the encrypted version of his or her message and be able to decrypt all packets using the same IV because he or she knew the original contents.

The second attack enables an attacker to construct and send a valid new packet using an existing packet. If an attacker knows the plaintext for an encrypted message, he or she can construct a new message, recalculate the ICV, flip the bits on the original encrypted message (changing it to the desired message), and have it received as valid. This is possible because although the ICV is part of the encrypted portion of the message, the CRC-32 is linear, meaning that it is possible to flip a bit in the message and determine the necessary adjustments to produce a correct ICV for the adjusted message.

The third attack is actually an extension of the second. The attacker tries to guess the information contained within the header, specifically the IP address of the destination. It is then possible for the attacker to flip the appropriate bits to send the packet to a controlled computer. The message will then be received and decrypted by the AP and forwarded unencrypted to the attacker, revealing the plaintext.

The last attack is actually a culmination of the previously mentioned attacks. Using these techniques, an attacker can build a table of IVs and the corresponding key streams. Once the table is built, all intercepted traffic can be decrypted. Given the short amount of time required to exhaust the IV space (two to five hours) and the relatively small amount of space required to store such a table (approximately 15GB), it does not seem that unlikely that there is a hacker out there somewhere with enough motivation to accomplish this.

In response to the UC Berkeley study and the resulting media frenzy, the *Wireless Ethernet Compatibility Alliance* (WECA) published its response to reiterate its vision of the WLAN security and to describe the current state of WEP security. Although WECA acknowledged the validity of the UC Berkeley paper, it maintained that the attacks were sophisticated and appeared to require considerable time and effort to mount, and that WEP is an effective deterrent against most hackers. WECA also issued a reminder that WEP encryption was never intended to be an all-encompassing solution; where deemed necessary, other products should be incorporated in addition to WEP to ensure the protection of sensitive data. WECA also pointed out that the attacks described by the UC Berkeley team were known by the IEEE Security Task Group I by October 2000 and enhancements had been proposed and adopted as part of the IEEE 802.11i draft proposal.

A study published in March 2001 by the University of Maryland looked at flaws in APs based on the ORiNOCO/Lucent 802.11 Wave-LAN *Personal Computer Memory Card International Association* (PCMCIA) cards. The authors of this paper also outline an eavesdropping technique to defeat the shared key authentication mechanism of 802.11. The authors state that there are two flaws in virtually all APs based on the IEEE 802.11 standard. The first is that even with WEP enabled, the management messages are broadcast in cleartext. Information such as the SSID lies within certain management messages. As a result, attackers can easily sniff this information out. The second is that MAC addresses are susceptible to the same sniffing as the SSID because they must appear in cleartext even when WEP is enabled. Also, the authors pointed out that most vendors allow MAC addresses to be changed via software. This could enable an attacker to obtain the permitted MAC address, and then add the desired address into the wireless card.

The University of Maryland study also outlines a passive attack that can be accomplished by eavesdropping on the mutual authentication messages and, if WEP is enabled, using previously published WEP attacks. The authentication process consists of four messages that all use the same frame format. Therefore, if the attacker captures the second message containing the cleartext challenge and the third message containing the encrypted challenge, he or she now has the challenge in both encrypted and plaintext format along with the IV, and he or she can derive the key stream. Once that is accomplished, he or she can request authentication with the AP and respond appropriately to the challenge text. Note that the attacker does not need to know the secret key to accomplish this authentication. The passive attack just described works when WEP is not enabled for authentication. If WEP is enabled, the WEP attacks must also be mounted.

The most egregious flaws to date were published by Fluhrer, Mantin, and Shamir in July 2001. Their study describes an attack where the attacker can calculate the WEP key by using the known IV, which is transmitted in the clear, and guessing the first byte of the plaintext. Using the knowledge that the packets appended to each IP and *Address Resolution Protocol* (ARP) packet are identical and knowing the IV and the first bytes of the plaintext, the attacker can easily determine the secret WEP key. This attack is impossible to detect because it is completely passive and it also extends linearly.

Although the concepts outlined in the Fluhrer, Mantin, and Shamir study were published only as a mathematical proof, a team from AT&T Labs and Rice University were able to put the theory into practice and prove the attack to be feasible. Their results were published in a paper in August 2001. This attack was successfully accomplished using off-the-shelf hardware and software to capture the WEP keys from a network within only a couple of hours. Although the AT&T team did not release its attack code, other tools have since been made available. Two of the most widely known tools to implement this attack are AirSnort and WEPCrack and are available online.

Doc, It Hurts When I Do This— Enhancing WLAN Security

So, at this point, you are probably saying to yourself, "Let's just ban wireless networking until it becomes more secure." That may not be a bad idea—that is, if you can keep Fred in sales from running down to the local Radio Shack and installing his own network without telling you. So, what can you do to protect your company's data while keeping Fred and others happy and productive?

Implementing multiple layers of security is known as *defense in depth*. The theory behind this is that the more lines of defense you implement, the less likely a hacker will be to penetrate the system, and the more likely it is that any attempt will be detected. It is also more likely that if a system is protected by multiple layers of defense, a hacker will simply give up and move onto another target. Many people think of a firewall as the main line of defense, but having only a firewall is akin to the Maginot line and it may not hold. Adding other technologies such as intrusion detection systems, enhanced authentication mechanisms, and encryption would greatly enhance the security of the system. The following are some ways to protect your system. These methodologies are categorized into the four broad security risks we discussed at the beginning of this chapter and can be applied fairly equally to both wireless and wired networks.

Loss of Equipment

Even if you do not currently have a wireless network installed, you likely have remote workers or workers who travel with their laptop computers. Even if your entire workforce is not mobile, they probably leave their laptops in their office at night, thereby making them vulnerable to other workers, maintenance crews, or burglars. There are several steps that workers should be aware of and encouraged or required to do in order to protect the company's equipment, and ultimately the corporate data. Laptops and *personal digital assistants* (PDAs) should be locked in the desk drawer when leaving the office. Laptops can be secured with a cable. The company's name and ID should be permanently engraved on the equipment. Keep the make, model, and serial number of the equipment separate from the computer. Do not store passwords on the computer or on a piece of paper in the computer bag. There are also some high-tech technologies such as proximity cards that shut down the computer when the user is out of range.

All employees should be made aware of the vulnerabilities and be required to take the necessary precautions to secure both the equipment and the data it can lead to. The nature of mobile equipment makes that equipment appealing to mobile workers and susceptible to theft.

Virus Protection

Antivirus software can protect a system from known attacks or viruses. Malicious code can come in three general forms: viruses, worms, or Trojan horses. It can easily bypass perimeter defenses by disguising itself as legitimate network traffic. A *virus* is a program designed to self-replicate and to be executed without user knowledge. A *worm* is a program that is designed to replicate itself and spread across a network infecting all connected machines. A *Trojan horse* is a program that appears to be a useful program, but actually captures information or enables the hacker to control your machine remotely. The newest threats are from Java applets and active X controls. These are small applications that run within your web browser without your knowledge. They are supposed to be safe; however, several security holes have been reported. Even if you do not have a wireless network, most, if not all, of your user community probably uses the Internet in some way while connected to your corporate network, making the network vulnerable to these threats.

The hacker community is never idle for long; it is always looking for new ways to wreak havoc. That is why antivirus software should be updated frequently to protect against the newest threats known. Some antivirus software provides for automatic updates, which makes the task easier and requires compliance. If a virus is detected, the software can rid the system of the virus and, in some instances, correct any problems the virus created.

Misuse by Authorized Personnel

Implement Security Policies   A security policy is the first layer of implementing defense in depth. In order to have adequate security, a company cannot simply install a few products and forget about the topic. Security is a process and must be monitored, and the balance between functionality and security must be constantly weighed. An organization must decide what risks are acceptable and what risks are not acceptable. Once these decisions have been made, risk mitigation strategies can be evaluated. Security mitigation strategies include automated programs and manual maintenance, monitoring, and follow up.

These decisions on risks and risk mitigation strategies should be written into a set of policies and procedures. A *policy* is a higher-level document, which clearly states management's viewpoints on securing the corporations data. *Procedures* are lower-level documents, which clearly articulate responsibilities. Procedures should include details to provide an audit trail and the appropriate steps to follow at the time of a potential or actual security breach. Because the organization itself is dynamic, and technologies and related security vulnerabilities change rapidly, revisiting the security procedures periodically is a prudent practice.

A well-defined security policy should be one of upper management's first priorities. Unless the policy is well defined, a secure corporate network infrastructure cannot be implemented. A well-defined security policy must then be implemented correctly in order to provide confidentiality, integrity, and the availability of systems and data.

Security Awareness Assuming that upper management has clearly articulated the corporation's security policies and the IT professionals have correctly implemented them, the remaining step is to ensure that employees are aware of the security policies and understand their importance in maintaining corporate security. Employees actually cause more than half of all damage to computer networks, sometimes unintentionally and other times intentionally. Providing adequate training to employees concerning security policies ensures that employees are aware of their rights and responsibilities in maintaining a secure environment. This type of training will also warn disgruntled employees who may be tempted to damage the system that their actions are being monitored and appropriate action will be taken.

A security awareness program should focus on the following elements: Employees need to be educated on the potential threats to the company. Typically, people will think of hackers, but other threats include corporate spies, disgruntled or former employees, or simple mistakes made by honest employees. Many companies do not do a good job of disabling access to the system when an employee is terminated. This creates a huge opportunity for other employees to access the system using the former employee's ID or for the former employee to retain access to corporate data. Employees need to understand the assets that the company is protecting. These may include intellectual property, competitive information, personal information such as medical records, or physical technology such as new circuit board designs. Employees should be made aware of physical security. Although the use of access cards is becoming widespread, human nature makes it likely that a person who sees another person appearing to be an employee stranded outside a locked entryway will probably allow admittance to that person without question. Employees must know who to contact in case of a suspected security breach.

There has been quite a bit of press lately on identity theft, so employees should be cognizant that thieves sometimes retrieve documents from the trash in order to gain valuable information about a person or company. Employees should be sensitized to the proper way to dispose of sensitive documents. Smooth talking criminals can use trickery to entice employees to release sensitive information. The practice of getting an employee to disclose information to an unauthorized party is known as *social engineering*. It was used very successfully by one of the most infamous hackers, Kevin Mitnick, to carry out security breaches within a company. Many employees become annoyed at the many passwords, *personal identification numbers* (PINs), and other codes they are required to remember, so they keep written passwords under keyboards, in desk drawers, on sticky notes stuck to their telephones, or on personal bulletin boards. Others will rely on the computer to remember their passwords for them when offered. Training must be made available to ensure that employees understand the criticality of creating secure passwords, changing them on a regular basis, keeping them secure, and never sharing them with anyone.

Although using e-mail, intranets, and the Internet are valuable tools for conducting business today, employees should be made aware of the risks of using them. Although it is critical to educate all employees, the technical staff must be kept up-to-date on the latest security issues. Training must be available to the technical staff as well as seminars, bulletins, and other resources for maintaining competence. The technical staff must also realize that simply implementing the latest versions of detection systems is not enough—constant monitoring is also a necessity.

Attacks by Unauthorized Personnel

Use Built-In Features Although WEP is not the strongest encryption protocol around, it is better than nothing. By default, out of the box, WEP is turned off and many wireless networks are implemented with WEP disabled. WEP should be turned on and other security mechanisms should be added as the data warrants. Depending on the maintenance capabilities of your network administrators and your particular equipment, use MAC address filtering where feasible. Change the default SSID, but do not choose an SSID that uses your company name, street address, or the name of your products. Using information that easily identifies who you are makes you much more of a target. If your AP supports it, turn off the broadcast SSID feature. If you are implementing a wireless router, consider turning off the *Dynamic Host Control Protocol* (DHCP). By assigning static IP addresses, you can hinder IP address sniffing. Also consider changing the subnet from the default so that a hacker cannot simply guess the default 192.168.1.0 network and configure his or her client with an IP address in that range.

Change Default Passwords As with every system, not changing the default password supplied by the manufacturer is a major security risk in wireless. Default user IDs and passwords are widely known as they are published in vendor manuals and on the Internet. Many of these defaults have very powerful privileges assigned to them. Not changing the defaults also negates the audit trail; therefore, if a breach occurs, it is impossible to determine the specific user who caused it. When choosing passwords, keep in mind that users tend to choose passwords that are short and easy to remember. Unfortunately, this also makes them easy to crack. Using a three-letter password in all one case takes about 66 milliseconds to crack. Using a seven-character password with mixed case, including letters, numbers, and symbols, would take more than a year to crack. Corporate policy should dictate that all passwords be of sufficient length, contain letters, numbers, and characters and should not be reused. Corporate policy should also dictate that passwords be changed on a reasonably short periodic basis—for example, every 30 days.

Virtual Private Networks (VPNs) *Virtual private networks* (VPNs) can provide security in addition to that provided in the 802.11b standard. Many companies have existing installed VPNs as a way to provide remote users with secure access to the corporate network via the Internet. By implementing a VPN solution to secure wireless transmissions, the wireless network is treated as an untrusted network just like the Internet. The VPN servers act as a gateway to the corporate network by providing authentication and full encryption over the wireless network. The VPN provides a secure, dedicated path, or *tunnel*, over the wireless network. The user is then authenticated using a centralized authentication service such as *Remote Access Dial-In User Service* (RADIUS).

Four protocols can be used in a VPN solution. The *Layer 2 Forwarding Protocol* (L2F), the *Point-to-Point Tunneling Protocol* (PPTP), and the *Layer 2 Tunneling Protocol* (L2TP) are all designed to work at the link layer (layer 2). *IP Layer Security* (IPSec) is the fourth protocol and it is designed to work at layer 3. The L2F was developed by Cisco Systems and is similar to the PPTP, which was developed by Microsoft U.S. Robotics and several remote access vendor companies known collectively as the PPTP Forum. Recently, Microsoft and Cisco agreed to merge their respective protocols into a single, standard protocol called L2TP. L2F and PPTP use somewhat weak authentication and encryption algorithms. L2TP is derived from these two; however, it is designed to use other more secure methods of encryption and authentication. IPSec is usually considered to be the best protocol for VPNs in an IP environment. It includes the strongest encryption and authentication methods. However, because PPTP, L2F, and L2TP function at layer 2, they can handle traffic other than IP, such as the *NetBIOS Extended User Interface* (NetBEUI) and *Internetwork Packet Exchange* (IPX).

When installing a VPN solution, keep in mind that the traffic is only encrypted within the tunnel. A VPN that terminates within the corporate network is far less secure than one that terminates within a firewall. Terminating the VPN within a firewall will allow policy- or user-based tunnels, which provide a higher degree of security.

Install Firewall Software on the Network   A firewall lets you define and enforce access control rules between two networks. Most corporations have installed firewalls to restrict traffic between the Internet and corporate LAN. The firewall can be configured either to allow or prohibit certain traffic on the network. A firewall should also be installed between the wireless network and corporate LAN in order to control traffic and isolate any access breach to the wireless network.

There are two basic types of firewalls. *Filtering firewalls* monitor data packets on arrival and apply the rules configured to determine if the traffic should be allowed to pass or not. A *proxy server* monitors outgoing traffic. Users on the corporate LAN log onto the Internet through a proxy server. Although firewalls are very effective perimeter defenses, they cannot control data entering via portable media such as a diskette or CD-ROM.

Install Personal Firewall Software   A personal firewall enables you to control traffic entering and exiting your PC. You can define the rules determining what is acceptable traffic and what traffic is high risk. The firewall enables you to monitor data and issue alerts if any unauthorized access attempts are detected. Some personal firewall products are Sygate Personal Firewall, Network ICE's BlackICE Defender, Symantec's Norton Personal Firewall, Tiny Personal Firewall, and Zone Labs' ZoneAlarm Pro. Most of these personal firewalls include an intrusion detection system to enable you to see who is attempting access to your system. In order to make a decision on which firewall would be best for your purposes, determine if the firewall software will block outbound access to the Internet from spy software. Most firewalls block ports from inbound access, but do not all block outbound access.

Segregate the Data   Sharing files is one of the purposes of implementing a network in the first place, but regardless of what you learned in kindergarten, don't share any more than you have to. Segregate the data based on sensitivity and grant access on a need-to-know basis. Put system-critical data on a separate drive—preferably a separate physical drive, but at least a separate logical drive. Grant access only to the files or directories necessary rather than to entire drives.

Who Goes There?—Intrusion Detection Programs   If Fred in sales were to implement his own wireless network right out of the box, how would you find out about it? This situation is similar to the situation where an employee places an unsecured modem on his or her PC, bypassing perimeter defenses and allowing dial-in access to the corporate network. Intrusion detection systems must be put in place to monitor each segment of the wireless network to notify you if an unauthorized AP has been inserted into the network or if an unauthorized client has gained access. By routinely running wireless sniffer equipment, the IT staff can locate unauthorized stations. Some vendors include facilities to help detect unauthorized APs or identify attacks. If yours doesn't, a number of these products exist on the market including MobileManager from Wavelink, Sniffer Wireless from Network Associates, and AiroPeek from WildPackets.

Future Enhancements

802.1x

802.1x, a standard that was approved by the IEEE in June 2001, includes a new protocol—the *Extensible Authentication Protocol* (EAP). This protocol is used to authenticate the association between the client and AP. EAP is an encapsulation protocol, which enables various authentication protocols such as RADIUS, Kerberos, or smart cards to be implemented. 802.1x also provides for the secure delivery of session keys. WEP keys can be dynamically generated and distributed, eliminating many of the manual processes needed to reset shared WEP keys. When each client has its own distinct WEP key, carrying out the Fluhrer, Mantin, and Shamir attack will be much more difficult. In addition, many APs will also enable configurations to automatically refresh client WEP keys periodically.

As of the end of 2001, 802.1x products were just beginning to be introduced. Microsoft Windows XP was the only operating system supporting this standard. Agere, Cisco, and Enterasys have wireless cards and APs that support the protocol. Microsoft IAS, Funk Steel-Belted RADIUS, and Interlind RAD-P authentication servers support EAP.

Currently, WEP is the only encryption method supported, but the *Advanced Encryption Standard* (AES) is being considered. AES uses the Rijndael algorithm, which is a block cipher. The current encryption method uses the RC4 algorithm, which is a stream cipher. Block ciphers encrypt blocks of data at once, rather than doing it linearly. This makes it much more difficult to carry out attacks such as those outlined by the UC Berkeley team.

Whether your organization chooses to install WLAN technology will depend on a number of factors, including the benefits of utilizing the technology, the type of data to be accessed over the wireless media, government regulations such as the *Health Insurance Portability and Accountability Act of 1996* (HIPPA), and the organization's ability to implement risk mitigation strategies. With a well-thought-out implementation and a full understanding of the benefits and risks, wireless networking can play a role in most organizations. Given the current heightened awareness of security, taking a second look at your wired network and the organization as a whole would be prudent even if wireless networking is not in your future.

Implementation

Setting up a basic 802.11b wireless network can be relatively simple. In fact, the use of Wi-Fi-certified equipment makes it likely that it will work right out of the box. However, working and working optimally are very different things. For most corporate or industrial installations, a number of factors must be considered to implement a network that will meet performance standards and work well with any established communications infrastructure.

The design and implementation of a *wireless local area network* (WLAN) is a function of the physical environment and the intended use of the network. Although desired functionality imposes demands on the physical configuration, the physical environment imposes limitations on the use. The first step in solving this equation is to define the basic physical usage requirements. You need to determine what areas need to be covered and what areas you don't want covered (like the lobby or the company on the floor above). Is coverage required in a single building, multiple buildings, and/or the open space between buildings?

Once the "where" has been clearly defined, a site survey can be undertaken. The site survey should include items that affect the propagation of the radio signal such as the building materials and potential sources of interference. It should also identify areas where radio signals are prohibited or not desired. For example, a hospital may have strict regulations on the allowable radio transmissions. Nonradio-related items must also be considered. *Access points* (APs) and bridges require connectivity to each other (the *distribution system* [DS] in an infrastructure *basic service set* [BSS]) and possibly to external networks such as corporate LANs or the Internet. Other nonradio items include the availability of electrical power and aesthetic factors such as the visibility of equipment.

After the physical environment is known, the usage requirements must be established. This includes the overall number of users, the number of simultaneous users, and what areas have a high concentration of simultaneous user access (for example, conference rooms). With data rates as a function of distance, transmission power, and interference, determining performance expectations is key to choosing the proper equipment and physical placement. The users' mobility profiles (stationary, portable, or roaming) affect network addressing, mobility management strategy, and equipment purchasing options. For example, if users require connectivity to public WLANs as well as to the private system, the use of proprietary features may be limited. If traveling users require access to corporate LANs located in various offices around the country or around the world, consistency between the various corporate LANs will affect the success of the implementation. The security implementation is dependent on the applications and the types of user devices required (such as computers, *personal digital assistants* [PDAs], and limited input scanning devices). The site survey and the usage requirements study determine the required number of APs and WLAN bridges, where they should be placed, the channel selections, the antenna types, the optimum power levels, and the required security and mobility platforms.

Once the environment and usage requirements have been established, equipment selection and purchasing come into play. Some vendors offer products with additional features that are options within the 802.11b standard such as 128-bit encryption. Some vendors have gone a step further by adding proprietary features such as additional security and authentication features, key management, roaming, or *power over Ethernet* (POE). Other vendors offer separate platforms for mobility management and enhanced security. A decision must be made whether to use proprietary features or features that are provided in the standard. Using proprietary features usually requires that all wireless equipment be from the same vendor. When making the decision of whether to use these proprietary extensions, all environments that may need to be supported, such as the home office or public WLANs found in airports and hotels, need to be considered.

Infrastructure

Probably the most likely reason for installing a WLAN is to provide mobile access to the existing corporate network. If this is the case, understanding the implications of the current infrastructure is one of the first tasks to undertake. The *Transmission Control Protocol/Internet Protocol* (TCP/IP) is the most widely implemented network protocol; however, others such as *Sequenced Packet Exchange/Internetwork Packet Exchange* (SPX/IPX), *NetBIOS Extended User Interface* (NetBEUI), LANtastic, or Pathworks are also frequently implemented. Support for each of these protocols is dependent on the wireless vendor. By installing the appropriate wireless driver, the network stack will not detect that it is running on a wireless connection rather than a wired connection.

The topology of a WLAN is much different than a wired LAN. The signal range will limit connectivity; therefore, complete coverage may not be achieved. This fact will break some assumptions made by the higher layers so the network must either be divided into cells that are managed by an AP or forwarding at the *Medium Access Layer* (MAC) level must be implemented. Each AP is the center of a cell and will operate on its own channel. The AP acts as a bridge connecting the wireless network to the wired network and relays packets between the two. The AP bridge parses the MAC headers to filter the data and changes the MAC headers depending on the protocol being used.

Most APs also provide access control to allow or disallow access to the network, roaming, and out-of-range forwarding. In order to implement these features, all the APs in the coverage area must be connected on the same IP subnet of the wired segment. Every station must associate with one and only one of the APs to avoid confusion. The station will associate with the AP that has the strongest signal, which may or may not be the physically closest AP. In order to implement roaming, when a station moves, it will automatically reassociate with another AP in order to maintain its connection to the wired LAN. Out-of-range forwarding refers to the process by which a station communicates to another station that is not in its coverage area. The AP that the sending station is associated with forwards the packet through the wired network to the AP where the destination is associated. Although it is true that the clients (laptops) do not require wired connections, the configuration must include cabling to connect the existing network to the APs, which is the portal between the wired and the wireless network. How easy or difficult it is to get CAT-5 cabling to the APs will have a direct effect on the overall cost of the implementation, the *return on investment* (ROI) of the project, and the viability of some implementation options. Once cabling has been established to create a portal, secondary APs can be implemented as a *wireless distribution system* (WDS).

How you get power to the APs is another important consideration. If the site has restrictions for where AC power outlets can be installed, these restrictions may limit the success of the WLAN implementation. One option provided by a few Ethernet switches is to enable power transmission over CAT-5 cabling, which is called POE. Some vendors provide add-on hardware that enables power to be distributed down existing cabling to the AP. Using POE may lower the overall cost of the implementation by eliminating the need to extend AC power to each AP, but be aware that if this option is utilized, not only is it an extension of the 802.11 standard, but it also makes utilizing a WDS impossible.

Site Survey

A site survey involves scouting the physical layout and properties of your implementation site. A good place to start is with the building blueprints. Most vendors supply site survey tools with their hardware. By utilizing these tools on a laptop equipped with a wireless *network interface card* (NIC), you can determine which areas may have a weak signal strength or poor signal quality. Your site survey should also include an investigation into potential causes of *radio frequency* (RF) interference. Survey people in the facility to try to identify any devices in use that may cause undue interference. There are also techniques and tools on the market to help you to identify the sources of interference. These tools are discussed in more detail in the section "Radio Interference."

AP Placement

Implementing the hardware that an 802.11 WLAN is made up of is relatively simple. At each node, a wireless network card and the included appropriate drivers must be installed. The implementation gets a little trickier when determining where to install the APs. The 2.4 GHz frequency band in which 802.11b wireless networks operate contains 80 MHz of spectrum. Typically, each *direct sequence spread spectrum* (DSSS) channel used by 802.11b takes 22 MHz of bandwidth, although it can vary. Interference between channels is minimized by using a minimum of 25 MHz of bandwidth for each channel. Note that interference is minimized, but not completely eliminated. Receiver design including antenna diversity, which is discussed in the section "Antenna Selection," will further minimize interference. Therefore, by using a minimum of 25 MHz of bandwidth over the 80 MHz of spectrum, three APs with noninterfering channels can be located in an overlapping coverage area.

Each AP provides 11 Mbps of bandwidth, so placing additional APs next to the original AP enables the bandwidth to be scaled from 11 to 33 Mbps. This solution will not provide an individual user with 33 Mbps of bandwidth, but it will provide this to be shared among all the users in the area. In 802.11b networks, each user associates with one AP at a time; therefore, a user must share the bandwidth of that AP with all the other users currently associated with the same AP. Installing more APs in an area provides the additional capacity to support more users in a denser population since fewer users share a given AP.

Changing the size of the coverage areas can also scale capacity and bandwidth. The transmission power of most APs ranges from 1 mw up to 100 mW. The range of the radio signal is obviously affected by the transmission power. The higher the transmission power, the longer the range of the signal. If you want to install your network in an office setting with cube walls and open space, a higher transmission power is appropriate. If your network is to be installed in a small office, conference room, or test lab, lower transmission rates may be more appropriate. Using lower power settings also enables more APs to be installed in a given area than if a higher transmission power is used. Instead of installing three APs transmitting at 100 mw and providing a total aggregate bandwidth of 33 Mbps in a given area, additional APs could be installed at a lower power to provide a greater total aggregate bandwidth.

One other important consideration is to ensure that roaming users have adequate coverage without installing unnecessary and expensive APs. By overlapping coverage areas, you can ensure that users do not experience any gaps in coverage. If you will be using wireless technology in an environment such as a warehouse where the devices are installed on rapidly moving equipment, the reassociation necessitated by the equipment traveling between APs makes it necessary to increase the density of AP placement. You also need to be aware of the security restrictions involved in this decision.

Channel Selection

Most APs come configured with a default channel that determines which part of the spectrum they will use. When installing the AP, the radio channel of neighboring APs will need to be reconfigured to use different nonoverlapping channels. Remember that because radio signals propagate three dimensionally, neighboring APs can be located on floors above or below your installation or even within your neighbor's installation.

By installing multiple APs on the same frequency, range and fault tolerance can be increased; however, this will not increase overall bandwidth. If one AP in a segment fails, wireless clients can seamlessly roam to a different AP with no interruption in service. By installing three clients on different channels, performance can be improved. Each wireless channel provides approximately 5 Mbps of throughput. Most clients will not allow association to be forced to a particular AP. Therefore, to achieve the most improved throughput rates, multiple overlapping cells with different frequencies must be built. Usually, the AP that a client will associate with is based on signal strength. One exception is 3Com, which considers both signal strength and load when associating to an AP.

Performance Factors

The effective range of your wireless network equipment can be affected by a number of factors. The propagation of radio waves is affected by environmental conditions. The frequency of the wave also affects the propagation of the wave. For example, low-frequency waves usually travel through plaster walls, whereas higher-frequency waves bounce off the same wall, losing energy in the process. The interaction of the radio waves with the environment configuration causes the signals to vary or fade. Additionally, the environment is not static. People may be moving, other objects in the environment may be moving, and the nodes themselves may be moving. These movements may result in changes in the signal. The signal between two nodes may alternate between good and poor on a per-packet basis. These fading factors will cause transmission errors that the system must be able to recover from, which in turn adds overhead to the system.

Path Loss and Multipath Loss

Usually, in a wireless environment, there is not a direct line of sight between the transmitting and receiving stations. This results in what is known as *path loss* or *multipath loss*. Path loss is the loss of signal power between the AP and client as the distance between them increases. Obstacles such as walls, ceilings, trees, fish tanks, furniture, and tinted glass can affect path loss. Transmission frequency can also affect path loss. Generally, higher frequencies cannot achieve as great a transmission as lower frequencies. Radio signals bounce off objects within the environment such as walls and furniture, which means that the signal can take multiple paths to its destination. When this occurs, the signal received may come from different directions with different strengths and the receiver must combine these reflections to arrive at the originally sent message.

The main problem with multipath is that it creates delay spread. Depending on the number of reflections and the speed differences between signals, the signals arrive at slightly different times at the receiver, creating an echo effect. This can create destructive interference when the echo of the previous signal corrupts the current signal. Because the various parts of the signal vary in amplitude and phase, the resulting rebuilt signal may be combined in a manner that severely degrades the signal's strength. The AP must properly demodulate and decode the signal into the originally sent transmission. Current DSSS products use techniques such as diversity antennas as well as signal-filtering and decision-making software to select the better signal to address the multipath problem.

Radio Interference

Given that wireless networks use RFs in the unlicensed *Industrial, Scientific, and Medical* (ISM) band, your implementation must anticipate and counter interference from other users in the band. Radio interference consists of the presence of unwanted signals that disrupt normal system operations. Interference is caused by other sources that also operate using the 2.4 GHz frequency such as microwave ovens, baby monitors, cordless phones, stadium lights, other WLANs, Bluetooth devices, and anything else that operates in the unlicensed 900 MHz or 2.4 GHz frequency band. Radio interference can result in static or no signal at all.

The 802.11 MAC protocol will recognize a sufficiently strong interfering signal as an 802.11 station transmitting a packet. The protocol then causes 802.11 stations on the WLAN to wait for an indefinite period of time until the interfering signal ceases transmission. Conversely, interfering signals will not likely follow the 802.11 protocols, so the interfering signal can begin transmitting while an 802.11 WLAN station is transmitting a packet. When this occurs, the receiving stations will receive the packet with errors; therefore, it will not transmit an *acknowledgement* (ACK), forcing the original station to attempt to retransmit the packet. All of this will add overhead to the network, causing delays and performance degradation. In some cases, 802.11 stations will automatically switch to a lower data rate, or, in some uncommon cases, wait until the interfering signal completely ceases.

Unfortunately, when installing a WLAN, you cannot "see" this interference and this lack of visibility impairs your ability to determine where to place your APs. By performing an RF survey, you can gain the knowledge of where interference is a problem and plan for adequate coverage and protection from interference. Traditionally, a site survey is performed by carrying around a laptop and measuring the signal strength and quality at various points throughout the site. This helps determine the coverage area between end users and APs. Typically, this would be done after making a best guess as to where the APs should be placed. If during testing, the coverage is not what was expected, the AP can be repositioned and the test can be repeated until adequate coverage is obtained.

This traditional method is fairly inexpensive, but lacks the capability to detect interference that occurs from other RF sources or between APs. By utilizing an 802.11b spectrum analyzer, you can visually see the amplitude of signals in any given channel. The 802.11b spectrum analyzer gives you the ability to isolate 802.11 signals from other sources and locate and eliminate the source of the interference. This type of tool also enables you to monitor channel usage and overlap. Remember that only three APs can operate in the same area on different channels, so by using these tools, you can better plan and implement the varying channels to achieve the greatest coverage.

If you are considering using a wireless phone in your home or office, consider choosing one that does not operate in the 2.4 GHz range. Microwave ovens should be operated more than 10 feet from an AP or station in order to prevent performance degradation. Bluetooth-enabled devices may also cause interference when operated in close proximity to 802.11 stations. The 802.11 and 802.15 standards groups are working on a standard to permit the coexistence of Bluetooth and 802.11 devices. 802.11 WLANs, such as those operating in neighboring companies, can cause interference unless you coordinate the selection of 802.11b channels with the other party.

Antenna Selection

Given that the 802.11b security features have been shown to have vulnerabilities, anyone who can hear the radio signals emanating from your WLAN can intercept them. However, properly implementing the antenna technology on a WLAN can reduce the emission of radio waves outside of your intended coverage area and make it much more difficult to eavesdrop. In addition, implementing a well-designed antenna will also make your WLAN less likely to suffer degradation in performance from interference by other users of the spectrum.

There are several types of antennas to choose from. An omnidirectional antenna receives in all directions. Directional antennas receive in an angle. The more narrow the angle, the less that unwanted signals and background noise are received. Directional antennas are offered as an option on some wireless equipment. Sectored antennas are similar to directional antennas. A set of directional antennas is assembled vertically, each one covering a section of the horizon. When communicating with a specific node, the sector covering the node is selected. This gives the benefit of directionality, but does not sacrifice coverage area. A dipole antenna, which is an omnidirectional antenna, is the most common WLAN antenna.

The radiation pattern of any antenna can be divided into two perpendicular planes known as the *azimuth* and *elevation*. The azimuth and elevation patterns are available for every commercial antenna. The axis of the dipole is known as the azimuth and the dipole radiates energy equally in all directions around the azimuth, but does not radiate above or below the wire, resulting in a donut pattern. Placing a dipole antenna in the center of the floor causes the radio waves to radiate along the floor with some radio waves reaching floors above and below the AP. In order to prevent the radio waves from leaving an intended floor and reaching the floors above and below, the radiation pattern must be attenuated. One way to accomplish this is to use an antenna that has several dipoles stacked on top of each other. This causes the antenna to radiate the signals in phase and concentrate the energy in a pattern that will resemble a squashed donut. In this way, the likelihood that an unintended recipient of the signal will be able to intercept the signal is reduced.

A dipole antenna transmits its signal so that the energy radiates parallel to the dipole and perpendicular to the floor. This is known as a *vertically polarized signal*. Turning the antenna 90 degrees causes the antenna to radiate in a horizontally polarized fashion. Depending on your implementation, one orientation may work better than the other.

In order to increase the transmission distance, concentrate the signal in the work area, and limit the amount escaping outside the building (or work area), a directional antenna can be used. A directional antenna radiates its signal in a beam, and very little energy is radiated to the back and sides of the antenna. There are several types of directional antennas. A *yagi* is a cylindrical directional antenna. A *patch antenna* is a flat antenna intended to be flush wall mounted. The patch antenna will radiate energy in a hemispherical pattern. Often simply called a *dish antenna*, a *parabolic dish antenna* is a concave antenna and usually provides the most targeted signal, which makes it useful in long-distance point-to-point implementations.

Another factor to consider when purchasing an antenna is the *standing wave ratio* (SWR). The SWR measures the amount of energy reflected back by the antenna. An SWR of 1:1 is perfect, meaning that there is no reflected energy, which could cause both signal strength degradation as well as spurious radiation from the cable to alert a radio signal sniffer. When buying an antenna, it is important to obtain and evaluate specifications and polar plots to determine the SWR ratio over all regions of the frequency range.

Most APs are equipped with two antennas in order to allow for antenna diversity. Antenna diversity is a means to overcome the fact that the channel signal can fade in and out over time. The conditions at each antenna, even though they may be only a few inches apart, may be very different at any given point in time. At any given moment, one antenna may have good reception and the other may have bad reception; at the next moment, the signal strength may be quite different at each antenna. One antenna is used as the primary transmission and reception port, and the other is polled prior to receiving each packet in order to determine which antenna has the best signal strength. By choosing the best antenna for each reception, the system can avoid most of the fade-out periods.

Roaming and Mobile IP

In order to enable users to maintain a continuous connection while walking around from one place to another, the network must let users switch between APs. Most WLAN vendors support this type of connectivity by having the stations automatically associate with the new AP as the station moves. The planning issue in the installation is determining how to divide the infrastructure into subnets. Installing APs in two different subnets requires traffic to cross a router, which many vendors do not currently support. The options are to either connect all the APs to the same subnet, which may require extra cabling, or to utilize Mobile IP.

If your network protocol is IP, you can implement the RFC 2002 Internet standard known as Mobile IP. IP addresses are designed to refer to a particular subnet so when a user connects to an AP outside of his or her subnet, IP traffic will not be able to find its way to the new location without some help. Mobile IP lets a user belong to a particular network, but have his or her mail forwarded. When the user moves to a different subnet, the user notifies the home subnet of his or her current location. The home subnet intercepts traffic intended for delivery to the mobile user and forwards it to a special node in the network known as the *foreign agent*. The foreign agent then forwards the message to the roaming user. Messages sent by the roaming user do not have to use this mechanism. These messages can travel directly to the recipient resulting in a triangular pattern of conversation.

Channel Access

The channel access mechanism is specified in the protocol and defines how the node uses the physical medium. In other words, it defines when to listen and when to transmit. The channel access mechanism used by most WLANs is *carrier sense multiple access with collision avoidance* (CSMA/CA). CSMA/CA that is used in wireless networks differs from *carrier sense multiple access with collision detection* (CSMA/CD) that is utilized in the wired network. A wired network adapter can listen to the cable to detect collisions as they happen. In a wireless environment, the same channel cannot be used to both transmit and receive because when a station is transmitting, its own signal will drown out any others arriving at the station, thereby preventing the station from listening to the network. In this environment, the MAC protocols must avoid collisions rather than detect them. By using CSMA/CA, the wireless network can ensure that collisions are minimized and that messages are reliably delivered.

By using CSMA/CD, the wired network can detect when two devices have transmitted simultaneously, causing a collision. When a collision is detected, each device waits a brief, random amount of time and then retransmits. Using CSMA/CD, a device on the wired network must listen on the cable to ensure that no other device is transmitting. If the cable is clear, the computer sends the smallest message possible (64 bytes) and then listens for a collision. If a collision is detected, a randomly generated number is used to determine a period of time that the computer must wait before attempting the transmission again. If no collision is detected, the computer continues transmitting the remainder of the message.

When collisions occur, the performance of CSMA/CD systems degrades rapidly. When these systems are implemented where keyboard input is the norm and data is sent in short, random transmissions, the systems work fairly well. However, the transmission of long files often creates delays. Implementing these systems with client/server systems increases the likelihood of collisions and the occurrence of collisions quickly rises, resulting in unacceptable performance.

Alternatively, the CSMA/CA protocol causes the wireless devices to wait until the air is CTS. The TCP network protocol was designed for wired LANS where packet losses are infrequent. When a packet is lost, TCP will assume that it was lost due to congestion, and in order to reduce the congestion, TCP will slow down. On the wireless medium, there are more errors because collisions cannot be detected, so TCP will believe that congestion has occurred and will slow down if a method for reducing errors is not implemented. By using CSMA/CA, the detection and elimination of packet losses is enhanced; therefore, the TCP protocol will not see the channel as congested and will not slow down.

This method, which is referred to as the *distributed coordination function* (DCF), prevents collisions by forcing each computer to announce its intention to transmit. CSMA/CA utilizes a random backoff time when the station detects that the WLAN is busy. A minimum time span between frames sent by a given user is chosen. After a frame has been transmitted from a station, that station must wait the amount of time dictated by the backoff time interval before the station can attempt to transmit again. If the channel remains busy, a second interval, which is less than the first, is implemented. This process continues until the time interval approaches zero and the station is allowed to transmit. This protocol will ensure that channels are shared among the stations and that collisions are avoided since the likelihood that two stations will randomly select the same backoff time interval is slim. Once the station is given permission to transmit, the device sends a *Ready to Send* (RTS) packet to the destination. If the destination receives the packet and is ready to receive, it sends back a *Clear to Send* (CTS).

This protocol works well because every node within range of the receiver hears the CTS even if it could not hear the RTS; therefore, it is notified that a transmission is occurring. Without this CTS being heard, these nodes could potentially cause collisions when the sending node is hidden. In the same way that these nodes may not have heard the intention to send, they may also not hear the actual transmission. To overcome this, the RTS and CTS messages contain a value indicating how long the transmission will last. This feature is known as *virtual carrier sense* and comprises the avoidance portion of the protocol.

The use of RTS and CTS packets also lowers the overhead involved if there is a collision because if two nodes attempt to transmit at the same time, only the RTS packets collide and the CTS is not received. In this manner, they will only lose the RTS message, not an entire message packet.

As mentioned previously, when packet losses occur, TCP performance can deteriorate. Most MAC protocols also implement positive ACK and MAC-level retransmissions to reduce the amount of errors and, in turn, the performance degradation inherent in the TCP protocol. The process is fairly simple. Each time a packet is received by a node, the node immediately responds with an ACK indicating that the packet has been successfully received. If the transmitting station does not receive the ACK, it knows that an error occurred and retransmits the packet. The MAC protocol does not transmit the next packet until the previous packet has been acknowledged as received. This will prevent the packets from arriving out of sequence, which TCP also does not deal with well. ACKs are embedded in the protocol so that they cannot cause collisions. Most new WLAN products include these features, but some older products may not. Wireless *wide area network* (WAN) protocols also do not include these features due to the length of the roundtrip delay to do so.

A couple of performance implications are contained within this scheme. First, the transmitting node must receive the ACK from the receiving station before it can process any other operation. When the wireless network's base station is the transmitting station, the entire network is slowed because all of the remote nodes must wait to send transmissions to the base station. Second, when there is a hidden node, two nodes may initiate the sending process at the same time, resulting in collisions and additional overhead. For example, this occurs when node X can communicate with node Y and node Y can communicate with node Z, but node X cannot communicate with node Z. Node X may sense that the network is clear, when actually node Z is transmitting to node Y. The CSMA/CA protocol will notify both nodes X and Y that they must wait before transmitting. Nodes X and Z may initiate this transmission process at the same time because they cannot detect each other's attempts directly. This is accomplished by using RTS/CTS handshaking; however, this handshaking adds significant overhead so it is not used for small packets or networks carrying light traffic loads.

An option in the 802.11b standard is the *point coordination function* (PCF). The PCF can be used for the prioritization and coordination of traffic. The AP can use PCF to assign priority to a client on a frame-by-frame basis. PCF is a very powerful mechanism because it can be used for applications that are sensitive to timing delays such as voice.

Application Configuration

The same applications that your users are currently familiar with on the wired LAN will also likely work on the WLAN. However, the speed of the network and the size of the application may need to be taken into consideration. Frequently used applications may be better located as a local copy rather than trying to load them from the server each time. This same consideration can be extended to utilities and frequently used data files.

Network Management

Two types of tools are available for monitoring and managing the network. Client tools enable the user to monitor the strength of his or her radio connection. Network tools are used by the IT staff to monitor and manage the wireless network. Most wireless products support basic monitoring tools based on the *Simple Network Management Protocol* (SNMP) such as HP Openview, but these tools pose certain problems and still leave some tasks to be done manually. The client tools used will depend on the NIC purchased. The NIC chosen should be Wi-Fi certified and should include easy-to-use client-side utilities. The tools should enable you to monitor the signal strength and quality, the link status, and network statistics. Information about configuration options such as *Service Set Identifiers* (SSIDs) and *Wired Equivalency Protocol* (WEP) keys should be available.

It is also important that the tools enable the user to maintain multiple profiles so that the he or she can switch between them as needed. Network tools provide both monitoring and management capabilities. Most APs include tools to manage the network. These tools typically enable the IT staff to set up the network by performing tasks such as configuring the device name, selecting the channels, setting the SSIDs, IP addressing, setting security parameters, and configuring Ethernet settings. Ongoing maintenance tasks typically include changing any of the initial parameters, upgrading firmware, rebooting the AP remotely, and analyzing network interfaces and client connections. When choosing an AP, take the tools and how they will be accessed into account. Tools can be accessed via a web browser, command-line interface, software utility, and serial console interface.

Monitoring tools are typically purchased separately from the AP and can monitor a variety of network devices. These tools can be used to provide real-time monitoring and alert the IT staff of critical events or network outages. These tools can also be used by the IT staff to store information that can be used in trend reporting.

Many factors affect the successful implementation of an 802.11 WLAN. Understanding what the users' needs and expectations are and what the organization hopes to accomplish with the implementation will set the requirements. With sufficient planning, the use of robust survey tools, and a careful consideration of interoperability, performance, and the network management needs, the right equipment can be chosen to ensure that users' needs and expectations are met. A wide variety of options is available to configure and tune the network in order to ensure quality, security, and ease of use.

10

Applications

No technology can exist in a vacuum; the real test of any new technology is its presence in the real world. As we have heard recently with the unveiling of *IT* or *Ginger*, the first question asked of most new technologies is "What would I ever use that for?" This chapter will attempt to answer that question for wireless networking. This chapter will outline some of the real-world uses of wireless technology, the reasons for implementing wireless rather than wired networks, and the benefits of the applications. Each of these implemented solutions is presented based on the type of application—retail, logistics, manufacturing, healthcare, education, government, and others.

This list is by no means exhaustive as new applications are being pioneered everyday. Specific vendors and their products are mentioned in these examples; however, they are not the only vendors or products available and we make no recommendations as to the reliability or applicability of the products to any particular environment.

Retail

Costco

The British subsidiary of the Costco warehouse chain sells approximately 3,500 products in each of its 125,000-square-foot outlet stores. Costco specifies the barcoding to be used on packaging to its vendors so it can ensure a consistent tracking mechanism from the vendor through the warehouse to the outlet store. The outlet stores operate as cash-and-carry membership warehouses, so when items are delivered, they are moved directly on pallets to the sales floor. Costco stock is rotated every six to eight weeks to provide unexpected items to its customers. In 1991, the company realized that it needed to find a standard platform to manage the systems and provide a path for future enhancements such as *voice over IP* (VoIP) and mobile electronic point of sale (ePOS). Costco implemented a Symbol Spectrum24® *wireless local area network* (WLAN) in its British stores. Costco says that it is gaining benefits due to the ease of management and reliability of the network. In fact, a network analyst for Costco says that in the 18 months that the network has been installed, he has received only two service request calls.

Chief Auto Parts

Chief Auto Parts is a privately held regional retail auto parts chain. The chain has experienced rapid growth with more than 550 stores currently operating in California, Texas, Nevada, Arizona, Arkansas, and Tennessee. Each store stocks more than 20,000 parts per store. Parts tracking and management, as well as enhanced customer service, motivated Chief Auto Parts to implement a Spectrum24 WLAN. Chief has two distribution centers, one in Texas and one in California. Each of the centers is between 245,000 to 300,000 square feet, and each processes a steady stream of deliveries from suppliers. Items are scanned with a Symbol LRT 3840 hand-held computer, which integrates a laser scanner and a Spectrum24 radio card. Receiving information is transmitted to one of several *access points* (APs) linked to multiple mobile *personal computers* (PCs) in the warehouse. From there, the data is sent via high-speed leased lines to the corporate AS400. It takes less than a second for a worker to record a receiving transaction.

Chief has also implemented a reverse distribution application where defective parts, recalls, and return-to-vendor used parts are scanned out of the system before they're loaded into the trucks to be returned. Chief also verifies each of the thousands of orders placed on the weekly shipments to each of the stores by scanning the barcodes prior to shipment.

Every Chief Auto Parts store also uses hand-held computers to perform inventory once a week. The parts are scanned and transmitted to the corporate computer via the WLAN. Store managers also use the scanners to perform shelf-label price verification, ordering, and returns.

The chief benefits of this technology are knowing what is in each store and using that knowledge to do target marketing and micromarketing. Through the use of wireless technology in the receiving function, information is collected instantaneously, eliminating the process of manual rekeying and unnecessary paperwork. This technology solution has enabled management to react quickly to the information and make better decisions. It has also cut the time required to perform routine tasks, enabling the store employees to spend more time with the customers.

FastFrog.com Case Study

Simon Property Group, the owner of 253 shopping malls in the United States, introduced a new way of shopping in December 1999 in a test market in Atlanta, Georgia. This concept gives customers the convenience of Internet shopping with the actual touch and feel of shopping at the mall and is known as yoursherpa.com. The Simon Property Group subsidiary clixnmortar.com plans to expand on the concept to leverage its relationship with its retailers and the 2.3 billion shopping visits a year made to its properties. Using Symbol's SPT 1740 pocket-sized computer, which includes a barcode scanner and wireless communications, shoppers can select items from different stores, but pay for all their items at one time. yoursherpa.com employees perform pickup and delivery.

Shoppers can also scan items at participating mall retailers using the Symbol CS 2000 scanner to personalize a wish list for posting at FastFrog.com so friends and family can shop easily for birthdays, weddings, anniversaries, Christmas, and other special occasions. Shoppers register at a mall kiosk, which is known as the *Pond*, choose a unique password, and check out a scanner called the *Zap-Stick*. Shoppers can then browse the participating stores to build their wish list, which is downloaded and posted to the web site when the ZapStick is returned to the Pond.

Sears

Sears Direct Delivery delivers appliances and other large items to customers. Sears needed to keep up with demand without building more warehouse space. The company implemented a WLAN in their warehouse to eliminate paper and streamline the process. The system uses a color touch-screen computer (Symbol's VRC 4000) that is mounted on each forklift. The Symbol Spectrum WLAN transmits data from a central server unit in the warehouse using thin-client software and displays each driver's pick list on the forklift-mounted unit. In a warehouse of 1,200,000 square feet that runs 24 hours a day, 6 days a week, the ability to make software changes at only one location is a huge efficiency gain.

CVS Corporation

The drug store industry competes on low prices and wide selection. The slim margins require stores to enhance productivity while enhancing customer satisfaction. Shoppers are generally in a hurry, want to find their desired items, and check out quickly. Maintaining the inventory displays and shelf price labels and reordering, receiving, and updating the current prices in the point of sale (POS) system are constant challenges. CVS implemented a wireless solution to manage these challenges. Store employees use the Symbol PDT 6840, which is a hand-held mobile computer with a built-in barcode scanner and WLAN connectivity. Data can be quickly transmitted to or from each store's host computer, and real-time item information and history can be maintained and accessed.

Dollar General

Dollar General is a retailer of basic consumable goods that operates over 2,800 stores and is growing at a rate of 400 stores per year. In order to support this growth, the chain implemented an integrated and automated warehouse system utilizing wireless technologies. In their Ardmore, Oklahoma, distribution center, the host computer is an IBM RS-6000 server. Intermec model RT 1740, RT 1750, and RT 1150 are used for wireless scanning. Purchase orders are generated on a server in Scottsville, Kentucky, and downloaded. Barcoded license plates are generated. Upon delivery, trucks are unloaded, cartons are cube-scanned and put on pallets, and the trucks are given license plates. The forklift operators are told where to put each carton, and as the operator scans the carton, inventory is automatically updated.

Store orders are downloaded from the Scottsville computer and items are picked the day the order is received. At the same time, work orders for restocking items picked for delivery are generated and labels are printed for the store cartons. By using this system, the distribution center can ship over 1 million cartons as compared to a peak day of 120,000 at a nonautomated center.

Logistics

Armellini Express Lines

Armellini Express Lines is a Florida-based transportation company that delivers more than half of all the fresh-cut flowers in the United States. For over 50 years, Armellini has received the boxed flowers from importers near the Miami International Airport and has taken them to its nearby terminal where they are sorted by destination and delivered to thousands of wholesalers. During a peak season, almost 100,000 boxes of flowers move through the Armellini terminal and are delivered by its 125 trucks.

With such a sensitive product, time and accuracy are essential. Mistakes such as loading flowers on the wrong truck or even delays are costly, so Armellini decided to invest in technology to reduce errors and speed the process. Barcodes containing the shipper and wholesaler's ID code, a unique box ID, and the contents are put on each box and scanned by checkers with Symbol LDT 3805 integrated scanning computers at the airport entry site. Billing is performed by *electronic data interexchange* (EDI) from the shippers to the Armellini mainframe in Palm City. When the flowers reach the distribution center, the boxes are received on the dock and scanned using a Symbol LRT 3800, which transmits the data in real time on the Symbol Spectrum One network. Receiving, staging, and loading information is matched to the initial billing information on the mainframe. Any discrepancies are identified in real time and corrected before the product leaves the warehouse.

UndercoverWear Lingerie Parties

UndercoverWear is a company that supplies lingerie for home parties. This company has been expanding rapidly. In 1997, the company implemented a new warehouse system using an 802.11-compliant wireless network from Symbol Technologies, mobile computers, and scanners. The company stocks more than 2,000 SKUs in only 400 square meters of warehouse space, so trolleys had to be custom-built for picking. Symbol LS5800 in-counter bar scanners are attached through the use of VRC 3940 vehicle mount terminals along with a Zebra A Series thermal printer. Two 12-volt gel batteries supply power to the trolley and only need to be charged on weekends. This entire system, which also includes a picking tote, is installed on a 500-mm-wide trolley.

Each item ordered is displayed along with its warehouse bin location. As the items are picked, they are scanned, and within a second, a label is printed showing both the party hostess and the party guest names. Completed orders are placed on a conveyor, packing and shipping documents are printed, and the package is weighed and shipped. The company has also recently installed wearable Symbol WSS 1040 scanners for hands-free receiving of stock from the delivery trucks.

Prior to the new system, one person was required to pick the order and a second person checked the order and completed the hostess and guest names. Training for pickers was a minimum of half a day and training for order checkers took more than two months. Since installing the system, training for order pickers is completed in minutes, and the second person required for order checking has been eliminated. Shipping times have been reduced from five days to the same day, back orders have been reduced, and the error rate has been reduced to virtually zero. The company has been able to maintain staff levels even though sales have increased 30 percent.

Manufacturing

Honeywell

Honeywell operates a 300,000-square-foot facility in Phoenix where they repair engine components for commercial, military, and executive aircraft. The accurate and complete capture of labor cost is extremely important in this production facility. If the labor component for the 800 employees who work in the different work centers is not captured properly, they must be charged to overhead and cannot be assigned to specific work orders. The facility previously used light pens on a wired network for gathering this data, but the wired system was unable to meet the demand for real-time information flow. The facility generates approximately 250,000 transactions a day and parts move frequently from workstation to workstation. With the previous batch system, many hours were spent looking for parts that were not where they were reported to be because a batch download had not yet occurred. Also, due to the limited data verification capabilities of the light pens, the workers could never be sure they had entered data correctly.

Honeywell implemented a new system using wireless technology from Intermec. With this system, workers use wireless Intermec 5020 hand-held computers loaded with a Web-based application and an Intermec Vista CCD scanner. The Intermec 5020 handhelds are used to move parts from location to location within the warehouse, providing real-time information, and because the Vista CCD scanners enable the barcodes to be scanned from 10 to 12 inches, the barcodes and scanners don't wear out as fast. Honeywell estimates that the system paid for itself in less than 90 days. Labor collection problems were reduced by 99.6 percent the first week.

Other employees at the Honeywell facility use the Intermec 6642 Pen Tablet computer with a touch-screen interface. The production control staff uses these devices to enable them to move around the floor and have the data to make decisions without having to return to their desks and use the old desktop computers. These devices are also used to conduct inventory. Previously, physical inventory counts were written down on paper and then keyed into the computer—a process that took two days. With the new scanners, the inventory count has been reduced to four hours.

Virco Manufacturing

Virco Manufacturing began its business in the 1950s as a local supplier of chairs and desks for schools in the Los Angeles area. They are now the largest manufacturer of educational furniture in the United States, employing over 2,400 people. Virco has over 100 Intermec EasyCoder 3400 printers connected to business systems throughout their manufacturing facilities for use in printing the UPC codes and "Care and Use" labels applied to their products. The printers are all connected to the same SAP system using a multidrop protocol, which uses the same cable and print server. The placement of the printers throughout the facility was restricted and the number of printers in each plant was limited to 32. Virco installed EasyLAN 10I networking cards so the printers can connect directly to SAP. This has eliminated the printer location restrictions and allowed Virco to move printers to where they are most needed.

HON Industries, Inc.

HON Industries is a big believer in rapid continuous improvement, so it is not uncommon to move production lines. The company had been using stationary barcode scanners that were hard-wired through the ceiling to a database that was hundreds of feet away. The system worked well, but it lacked the key flexibility needed. Moving a workstation could take anywhere from 6 to 20 hours using an outside company. Now that HON has upgraded to a wireless system, which includes Intermec's 5055 stationary, wireless, and touchscreen PCs, Intermec's ScanPlus™ 1800 Vista wireless hand-held barcode scanners, and Intermec 4400 barcode printers, production lines can be moved without moving cables. The benefits of the system include increased control over its production lines, shortened lead times, reduced operating costs, and customer satisfaction.

Employees can scan the color and pattern of the upholstery verifying that the correct combination of fabric, color, and pattern has been used. Once each custom-built chair is completed, the data is transmitted wirelessly to the database on the mainframe and updates the master production schedule. The data is then verified and a unique barcode is attached to the product as it is boxed to track the shipment from the manufacturing line to the warehouse, through the supply chain, and ultimately to the customer. When a customer calls to inquire about his or her order, the information can be accessed in real time and the customer knows the exact status. Customers can also order replacement parts using the unique product serial number.

Healthcare

The University of Michigan Hospital

The University of Michigan Hospital's Intensive Care Unit is a level-1 trauma unit and treats everything from trauma surgery to burn patients. When the unit is extremely busy, paperwork can get overlooked. In an effort to better capture clinical documentation and charges, the director of Trauma Services began looking for an automated solution. After evaluating the needs of the center, *Systems Engineering Consultants Inc.* (SEC) decided that three technologies were key—wireless communications, mobile computing, and handwriting recognition. The unit installed 10 wireless PCs, 20 Fujitsu Stylistic Pen Tablet computers with RangeLAN2 WLAN technology, and 25 wired desktop PCs. Doctors pick up a mobile unit containing all patient information to take with them on rounds. This system captures and stores patient information for immediate reference and also when the patient returns for subsequent visits. This has proven to be more effective than paper charts, which can often be misplaced. In a reverification audit by the American College of Surgeons, documentation compliance has become 100 percent. The director of Trauma Services also reports that revenue has increased 167 percent for the same patient population. The University of Michigan Medical Center is planning to use the system as a research tool and do chart reviews of patients in the Intensive Care Unit in minutes rather than weeks.

Children's Hospital of the Greenville Hospital System

Pediatric care is being brought to sick children who cannot get to the hospital either because of lack of transportation or because they cannot be moved. A van set up to house a mobile clinic brings care to young patients' homes. The pediatrician can conduct checkups, camp physicals, immunizations, and sick visits from the van. Making room for several hundred patient charts in the van has been a problem. The pediatricians don't know which records they will need so they try to have access to as many as possible. A solution that uses Range-LAN2 equipment from Proxim has been implemented.

A generator in the van is used to supply power to the MedServer Mobile server running multiuser Windows NT and Citrix Metaframe remote access software and CompuKID, a special program for pediatric medical records. The MedServer is placed on a desk in the van and plugged into an A/C outlet. A radio data connection between the file server and the hand-held computers is established, enabling doctors to review patient history and chart the current visit. Doctors use Sharp Mobilon TriPAd PV-6000 portable PCs to chart exams, immunization schedules, and vaccinations. Backup files are created by connecting the server to the wired network back at the hospital. One doctor reports a gain in gross revenue of 18 percent due to the efficiencies of charting patient care and a reduction in missed billing.

Multimedia Terminal Mobile (MTM) Project and the Steinbeis-Transferzentrum Medizinsche Informatik

The Steinbeis-Transferzentrum Medizinsche Informatik is a technology transfer center located in the Heidelberg Technology Park. The MTM Project stands for the *Multimedia Terminal Mobile*. This organization developed a portable and wireless *personal digital assistant* (PDA) that communicates through the *Universal Mobile Telecommunications Service* (UMTS) and software for Telemedicine, Telelearning, and Tourist Guides. Together, they have developed an application called the CHILI radiology project, which is making an effort to bring telemedicine to PDAs. The application runs on a Compaq iPAQ H3600 running Linux. Professional medical workers can view images, talk to each other using VoIP, and see each other's mouse pointers as they review cases to decide on an approach to treatment. The application running on the iPAQ provides access to a central server and performs two-way data synchronization between the iPAQ and the server as the users are teleconferencing. The system includes the capability to zoom, pan, and browse through patients, and it also allows the study of images and the entire series to give the teleconferencing participant the best possible overview.

Sweden's National Board of Health and Welfare

The Swedish government wanted to ensure that during a crisis the communications system would be in place in order to provide the best care possible. The project was called the *IS Swede* information system and uses hardware suppliers Symbol Technologies and Hewlett Packard to connect both hospital-based and central networks to the ambulance fleet in real time. Prior to implementation, the primary means of communication between the paramedics and the hospitals was CB radios and wide area cellular telephones. The IS Swede project implemented *Global Positioning System* (GPS) tracking, wide area networking, wireless local area networking, and mobile computing. The central servers and hospital networks store information on patient medical history, disaster planning, and other materials. These servers are accessible to hospital staff, ambulance services, healthcare officials, and the Sweden central emergency broadcast system through web interfaces. Each ambulance contains a Hogia vehicle computer used as a server and a wide area Mobitex radio that includes GPS tracking so the hospital can monitor the location of the ambulance. Paramedics also use a Symbols SPT 1749 802.11 wireless rugged Palm-based hand-held computer for data communications at an accident scene. Using the SPT 1740, paramedics can collect the pulse rate, blood pressure, overall patient condition, and medication administered at the scene. The information is stored on the device and sent to the Mobitex system. A WLAN-enabled Hewlitt Packard Jornada 820 is also connected to the Mobitex system and is used for transmitting more detailed information on injuries while on the way to the hospital. The Mobitex system also provides two-way data communications with the central network.

By using these technologies, communication is expedited and hospitals can be fully prepared with the right staff and material long before the ambulance arrives. The risk of errors is reduced because critical information at the scene is recorded by barcode, enabling paramedics to verify allergic reactions by accessing a person's medical profile in the central network.

Education

Saddle Ranch Elementary School

Saddle Ranch is a new elementary school in Highlands Ranch, Colorado. Saddle Ranch was designed to be a wireless campus. Schools in the Douglas County School District are networked using a T1 for both voice and data. Saddle Ranch's 31 iMac computers are equipped with the WaveLAN Ethernet converter, a small wireless AP that connects via an Ethernet connection. Other users use either WaveLAN cards in laptops or WaveLAN cards with *Industry Standard Architecture* (ISA) adapters in desktops to connect. Saddle Ranch also uses the new Macintosh ibooks with a built-in wireless card and antenna. There is also an iMac in each classroom for the students and teachers. Students can also check out the wireless ibooks for use in classrooms, libraries, and other common areas. The ibook has a six-hour battery so it can be used for an entire school day before recharging. Students can use the equipment to connect to the Internet or the school's instructional network. Teachers can use the equipment to access grades and learning materials. School administrators can also use the network for the myriad daily administrative tasks involved in running the school.

The Girls Preparatory School

Located in Chattanooga, Tennessee, the Girls Preparatory School has implemented a totally immersed wireless computing environment. The school, which has a liberal arts slant, wanted the girls to learn to use computers in everyday life and be prepared to enter today's technology-driven workforce. The school implemented a campuswide WLAN that used a Symbol Spectrum24 802.11-compliant infrastructure that was wired to the campus's web-serving backbone. The solution uses more than 30 APs in the ceilings and walls to cover five buildings and related public areas. Students can use their laptops from almost any campus location to wirelessly access the intranet, take tests on the internal testing sites, or use the Internet. Students use the system for e-mail, homework assignments, science experiments, research, web site development, and electronic study groups.

Government

City of Oakland

The City of Oakland has implemented a high-speed wireless system to boost the efficiency of its police force. The *radio frequency* (RF) infrastructure in place was not capable of handling the capacity necessary for transmitting large graphics files (like mugshots) in a timely manner. A system upgrade using the Total Roam Ellipse™ software by Padcom now enables patrol cars to receive large files via a WLAN available in parking-lot locations throughout the city. The Total Roam Ellipse software was used to create a *local area network/wide area network* (LAN/WAN) data network solution. When officers are in the 802.11 coverage areas, the LAN capabilities of the solution are automatically initiated, enabling officers to begin large file downloads. If officers must leave the service zone before completing the data download, the system marks their place and automatically resumes when they return to the coverage area. The city also chose to use Xcellenet's Afaria software to perform management tasks such as the central distribution of software applications and updates, tracking hardware and software assets, and information exchanges.

U.S. Marines

The *U.S. Marine Corps* (USMC) uses thousands of Symbol PDT 7240 mobile computers to track, manage, and deploy thousands of tons of supplies. As Marines are deployed, supplies and services are needed to supply and sustain the troops. Maintenance facilities, food, munitions warehouses, and package-handling systems must be in place to support the Marines anywhere they are sent across the globe. At major USMC logistics bases in the United States, the PDT 7240s are used to scan barcoded information, which is then communicated over a Spectrum24 wireless network for real-time inventory. In theaters of operation, inventory is tracked using the PDT 7240s and sent back to a wireless laptop mobile server. These devices are used to support the supply, transport, maintenance, inventory, and replenishment of equipment wherever Marines are sent.

The Town of Enfield, Connecticut

Founded in 1863, Enfield, Connecticut, is one of the oldest towns in the country. The local government uses the town network for many functions, including education, public safety and public works, tax assessment, planning and development, social services, attendance, payroll, insurance, and Internet access. When the town wanted to connect a school wing, a branch library, and an activity center to the town network, two of the driving factors were preserving the historic nature of the buildings and keeping the costs down. They chose

Enterasys's RoamAbout Network partially because of the lack of recurring monthly charges to maintain a T1 line and partly to avoid having to drill holes, knock down walls, or run cables in a building nominated for the national register.

The high school building was U-shaped and had a remote wing that needed network access. The distance was too far to use Category 5 cabling, and fiber would have cost approximately $8,000 and would have likely required an intermediate distribution facility. With the Enterasys RoamAbout Network wireless solution, the costs were approximately $2,500.

The library received a grant for Internet access and sharing computer resources. The library is located in a Carnegie building and was nominated for the National Historic Registry. By using the Enterasys RoamAbout network to connect to the LAN within the library, they avoided having to drill holes, knock down walls, or run cables in a historic building. This branch library is connected to the town network through a *virtual private network* (VPN) gateway.

Hospitality

Casinos

Several casinos including the Aladdin Hotel, Carson Valley Inn, and Mystic Lake employ wireless hand-held computers to give employees more freedom of movement around the casino while providing information on guests and machines. Hand-held Epson touch-screen computers communicate with the LAN server using Proxim RangeLAN2 WLAN adapters. Real-time information is maintained through the use of network adapters installed in the gaming machines. In order to play, the guests insert cards issued to them that contain information about them. The machines collect information about the time and amount of play, wins, losses, and even maintenance information. Employees can see on their wireless handheld the guest's name and his or her win/loss status, enabling the employees to call the guest by name and know who to give complimentary drinks, meals, or rooms.

Ice Hotel

The Ice Hotel located 125 miles north of the Arctic Circle near Jukkasjärvi, Sweden, opened in 1992. The hotel features guest accommodations and a bar made exclusively of blocks of sculpted ice and snow. The Ice Hotel is rebuilt with varying configurations each October because it starts to melt as the temperatures rise in the middle of April.

Although it would be possible to sandwich network cable between the 16-by-18-foot blocks of ice, there is a risk that the builders would cut the cable as they sculpted doors and furniture into the blocks. Also, because the hotel is reconfigured each year, the network would have to be replanned. Lastly, a cable would be necessary to connect the hotel to the nearby administration building, raising the risk that the network cable, which needs to lie on top of the frozen ground, would be cut by the snow-shoveling crew. Therefore, a wireless solution was designed; however, there were challenges. Have you ever tried installing an AP in an ice wall? Because the ice doesn't transmit the signal very well, antennae were installed on the outside of the buildings, but the hotel itself did not receive an AP. The handhelds used in the ice bar talk to the node in the closest building instead. Also, because the 2.4 GHz frequency is the same as that used by a microwave, the signal loses strength quickly as it expends energy to heat the ice. Although because the power level used (100 milliwatts) is much lower to that typically used by a microwave (1,000 or more watts), there is not much chance of ordering enough Bloody Marys to melt the ice bar walls.

Prior to the wireless system, waiters in the ice bar would take orders on paper and then give the slips to someone in reception who would enter the charges into the central computer. This method was impacted by the cold, damp climate in the Ice Hotel. The order slips would get moist, making credit-card impressions difficult. The ink in the pens sometimes solidified in the below-freezing temperatures. A new Symbol24 high-rate wireless network was installed and now the waiters connect to the network with a Symbol PPT 2746 Pocket PC to transmit the tab to the central computer. At check-in, each guest receives a barcoded ID card. When the guest orders drinks, the card is scanned to ensure the charge is billed to the correct bill. The handheld's Web browser connects to the hotel's intranet and a custom software application written by Frontec running on the central computer in reception updates the guest's bill.

The pilot project has been very successful; the hotel is considering expanding the wireless network so that handhelds can be used in other shops throughout the complex.

Opryland Hotel

With 2,883 guestrooms, 600,000 square feet of convention space, 9 acres of indoor gardens, a quarter-mile river, and over 4 million guests annually, the Opryland Hotel is the largest hotel-convention center under one roof in the world. Opryland, part of the Gaylord Entertainment Company, has implemented a solution that enables employees to check people in and out, process credit cards, print receipts, and program room keys from anywhere in or nearby the hotel.

The hotel is only five floors with the rooms and facilities spread out over a square mile. Due to the convention traffic, guest arrivals can peak just prior to the convention with up to 2,000 guests arriving virtually at the same time. Previously, there were 2 registration areas with 24 desktop computers. Hotel management felt that having guests wait in line at the registration desk for up to 15 minutes during registration was not making the right first impression.

Opryland chose LANSA partner IAD's Lodging Management System. Now, staff can register guests throughout the hotel using a hand-held Symbol SPT 1740 Web-enabled computer that fits into a pocket. The devices communicate using Symbol's Spectrum24 802.11b WLAN to APs located throughout the hotel to the iSeries running the core business system. Guests can be greeted curbside and checked in while their luggage is being unloaded, can be met at their room, or can be met at their first session of the convention for check in. Group check-in can also be done either in a restaurant or on a tour bus. Checkout can be handled in the same way, enabling a guest to check out during the last session of his or her convention.

Park City Mountain Resort

Park City Mountain Resort is located in Park City, Utah, deep in the Rocky Mountains. This area is notorious for cell phones not working, but the resort has successfully implemented a wireless network. Four ski-lift operators carry hand-held devices. Skiers ski up, wearing barcoded lift tickets. The lift operator scans the ticket with a hand-held device from Symbol Technologies communicating to the ticket office over a WLAN and can immediately validate the ticket. If the ticket is invalid, the ski patrol can escort the skier off the property. If the problem is that a season ticket holder has a problem with a credit card such as an expired card, the skier can give a new credit card to the lift operator and fix the problem immediately on the slopes. Skiers can also buy food, take lessons, rent equipment, and sign up for other activities using the same system.

The system is driven by line of sight, so all the radio repeaters and antennas at the lift are in the line of sight of the server room at the lodge. Much of the infrastructure was already in place, including a *Synchronous Optical Network* (SONET) provided by Qwest. Lucent's ORiNOCO™ APs were used to integrate the WLAN.

Jimmy Lu's Asian Bistro

In the restaurant business, it is widely known that customers are far more likely to complain about the service than the food. Jimmy Lu wanted to bring his waiters out from behind the POS system to interact with the customers. Mr. Lu integrated the Aloha Technologies TableService POS system with Ameranth Wireless's 21st Century Restaurant®. The 21st Century Restaurant system consists of three modules: Back Office Software for WLAN integration, Wait Staff Software for wireless ordering, and Payment Processing Software for wireless credit-card processing. Now waiters use a Windows-CE-based touch screen for taking and transmitting orders at the table. The system also enables waiters to check orders, split checks, review unavailable menu items, route orders to various kitchen teams, process payment, and print receipts all at the table.

This system has reduced training time, shortened dining time, and reduced the number of servers needed on the floor. Because the order does not need to be duplicated between the tables and the POS terminal, the waiters can manage larger sections of the floor and turn tables over more quickly, resulting in increased tips and higher employee satisfaction. Mr. Lu estimates that they have reduced table time by about 10 minutes, thereby reducing the wait time for a table and increasing the bottom line.

Other

ATP Tennis Tournaments

ATP Tours is the governing body of men's tennis. It was founded and is owned by tournaments and top professionals. There are more than 70 tournaments from January to November held on every continent, and the fans who follow the players number in the millions. ATP's objective is to promote growth in the popularity of the game. A top priority in this plan is to ensure that fans can get up-to-the-minute information on their favorite stars.

During tournaments, an online link to the palmtop PCs used by the umpires was essential. If cables were used, they could hinder players and others on the court, or if they were run high, they could obstruct the view of play. Wireless technology was an obvious solution, but it could not interfere with TV, radio, or other equipment. The solution also had to be portable and be able to be set up in different configurations and tested quickly at various buildings, complexes, and courts. Lucent Technologies (Agere) WaveLAN/IEEE was chosen. Umpires are equipped with Sharp HC-4100 palmtops with WaveLAN *Personal Computer Memory Card International Association* (PCMCIA) cards that are running software developed by Parisportinfo. Scores are transferred to laptops from the umpires' palmtops using WaveLAN WavePOINT base stations to the Web in less than 10 seconds.

At most tournaments, one or two WavePOINT units are needed to link to all umpires on all the courts. These are typically located on lighting towers, clubhouse roofs, and grandstands. The Wave-POINTS are connected to the laptops by fixed cabling. The laptops have *Integrated Services Digital Network* (ISDN) connections to the www.atptour.com server. The fans seem to love it; at the last Super 9 tournament in Rome, the server registered almost 3 million page views in a week.

APPENDIX A

FCC Regulations

Subpart A—General

Subpart B—Unintentional Radiators

Subpart C—Intentional Radiators

Radiated Emission Limits, Additional Provisions

Subpart D—Unlicensed Personal Communications Service Devices

Subpart E—Unlicensed National Information Infrastructure Devices

Subpart A—General

Section 15.1 Scope of this part.

(a) This Part sets out the regulations under which an intentional, unintentional, or incidental radiator may be operated without an individual license. It also contains the technical specifications, administrative requirements, and other conditions relating to the marketing of Part 15 devices.

(b) The operation of an intentional or unintentional radiator that is not in accordance with the regulations in this Part must be licensed pursuant to the provisions of Section 301 of the Communications Act of 1934, as amended, unless otherwise exempted from the licensing requirements elsewhere in this Chapter.

(c) Unless specifically exempted, the operation or marketing of an intentional or unintentional radiator that is not in compliance with the administrative and technical provisions in this Part, including prior Commission authorization or verification, as appropriate, is prohibited under Section 302 of the

Communications Act of 1934, as amended, and Subpart I of Part 2 of this Chapter. The equipment authorization and verification procedures are detailed in Subpart J of Part 2 of this Chapter.

Section 15.3 Definitions.

(a) **Auditory assistance device.** An intentional radiator used to provide auditory assistance to a handicapped person or persons. Such a device may be used for auricular training in an educational institution, for auditory assistance at places of public gatherings, such as a church, theater, or auditorium, and for auditory assistance to handicapped individuals, only, in other locations.

(b) **Biomedical telemetry device.** An intentional radiator used to transmit measurements of either human or animal biomedical phenomena to a receiver.

(c) **Cable input selector switch.** A transfer switch that is intended as a means to alternate between the reception of broadcast signals via connection to an antenna and the reception of cable television service.

(d) **Cable locating equipment.** An intentional radiator used intermittently by trained operators to locate buried cables, lines, pipes, and similar structures or elements. Operation entails coupling a radio frequency signal onto the cable, pipe, etc. and using a receiver to detect the location of that structure or element.

(e) **Cable system terminal device (CSTD).** A TV interface device that serves, as its primary function, to connect a cable system operated under Part 76 of this Chapter to a TV broadcast receiver or other subscriber premise equipment. Any device which functions as a CSTD in one of its operating modes must comply with the technical requirements for such devices when operating in that mode.

(f) **Carrier current system.** A system, or part of a system, that transmits radio frequency energy by conduction over the electric power lines. A carrier current system can be designed such that the signals are received by conduction directly from connection to the electric power lines (unintentional radiator) or the signals are received over-the-air due to radiation of the radio frequency signals from the electric power lines (intentional radiator).

(g) **CB receiver.** Any receiver that operates in the Personal Radio Services on frequencies allocated for Citizens Band (CB) Radio Service stations, as well as any receiver provided with a separate band specifically designed to receive the transmissions of CB stations in the Personal Radio Services.

This includes the following:

(1) a CB receiver sold as a separate unit of equipment;

(2) the receiver section of a CB transceiver;

(3) a converter to be used with any receiver for the purpose of receiving CB transmissions; and,

(4) a multiband receiver that includes a band labelled "CB" or "11-meter" in which such band can be separately selected, except that an Amateur Radio Service receiver that was manufactured prior to January 1, 1960, and which includes an 11-meter band shall not be considered to be a CB receiver.

(h) **Class A digital device.** A digital device that is marketed for use in a commercial, industrial, or business environment, exclusive of a device which is marketed for use by the general public or is intended to be used in the home.

(i) **Class B digital device.** A digital device that is marketed for use in a residential environment notwithstanding use in commercial, business and industrial environments. Examples of such devices include, but are not limited to, personal computers, calculators, and similar electronic devices that are marketed for use by the general public. Note: The responsible party may also qualify a device intended to be marketed in a commercial, business, or industrial environment as a Class B device, and in fact is encouraged to do so, provided the device complies with the technical specifications for a Class B digital device. In the event that a particular type of device has been found to repeatedly cause harmful interference to radio communications, the

Commission may classify such a digital device as a Class B digital device, regardless of its intended use.

(j) **Cordless telephone system.** A system consisting of two transceivers, one a base station that connects to the public switched telephone network and the other a mobile handset unit that communicates directly with the base station. Transmissions from the mobile unit are received by the base station and then placed on the public switched telephone network. Information received from the switched telephone network is transmitted by the base station to the mobile unit. Note: The Domestic Public Cellular Radio Telecommunications Service is considered to be part of the switched telephone network. In addition, intercom and paging operations are permitted provided these are not intended to be the primary modes of operation.

(k) **Digital device.** (Previously defined as a computing device.) An unintentional radiator (device or system) that generates and uses timing signals or pulses at a rate in excess of 9,000 pulses (cycles) per second and uses digital techniques, inclusive of telephone equipment that uses digital techniques or any device or system that generates and uses radio frequency energy for the purpose of performing data processing functions, such as electronic computations, operations, transformations, recording, filing, sorting, storage, retrieval, or transfer. A radio frequency device that is specifically subject to an emanation requirement in any other FCC Rule Part or an intentional radiator subject to Subpart C of this Part that contains a digital device is not subject to the standards for digital devices, provided the digital device is used only to enable operation of the radio frequency device and the digital device does not control additional functions or capabilities. Note: Computer terminals and peripherals that are intended to be connected to a computer are digital devices.

(l) **Field disturbance sensor.** A device that establishes a radio frequency field in its vicinity and detects changes in that field resulting from the movement of persons or objects within its range.

(m) **Harmful interference.** Any emission, radiation, or induction that endangers the functioning of a radio navigation service or of other safety services or seriously degrades, obstructs, or repeatedly interrupts a radiocommunications service operating in accordance with this Chapter.

(n) **Incidental radiator.** A device which generates radio frequency energy during the course of its operation although the device is not intentionally designed to generate or emit radio frequency energy. Examples of incidental radiators are dc motors, mechanical light switches, etc.

(o) **Intentional radiator.** A device which intentionally generates and emits radio frequency energy by radiation or induction.

(p) **Kit.** Any number of electronic parts, usually provided with a schematic diagram or printed circuit board, which, when assembled in accordance with instructions, results in a device subject to the regulations in this Part, even if additional parts of any type are required to complete assembly.

(q) **Perimeter protection system.** A field disturbance sensor which employs RF transmission lines as the radiating source. These RF transmission lines are installed in such a manner that allows the system to detect movement within the protected area.

(r) **Peripheral device.** An input/output unit of a system which feeds data into and/or receives data from the central processing unit of a digital device. Peripherals to a digital device include any device that is connected external to the digital device, any device internal to the digital device that connects the digital device to an external device by wire or cable, and any circuit board designed for interchangeable mounting, internally or externally, that increases the operating or processing speed of a digital device, e.g. "turbo" cards and "enhancement" boards. Examples of peripheral devices include terminals, printers, external floppy disk drives, and other data storage devices, video monitors, keyboards, interface boards, external memory expansion cards, and other input/output devices that may or may not contain digital circuitry. This definition does not include CPU boards, as defined in paragraph (bb) of this Section, even though a CPU board may connect to an external keyboard or other components.

(s) **Personal computer.**   An electronic computer that is marketed for use in the home, notwithstanding business applications. Such computers are considered Class B digital devices. Computers that use a standard TV receiver as a display device or meet all of the following conditions are considered examples of personal computers:

(1) Marketed through a retail outlet or direct mail order catalog.

(2) Notices of sale or advertisements are distributed or directed to the general public or hobbyist users rather than restricted to commercial users.

(3) Operates on a battery or 120 volt electrical supply.

If the responsible party can demonstrate that because of price or performance the computer is not suitable for residential or hobbyist use, it may request that the computer be considered to fall outside of the scope of this definition for personal computers.

(t) **Power line carrier systems.**   An unintentional radiator employed as a carrier current system used by an electric power utility entity on transmission lines for protective relaying, telemetry, etc. for general supervision of the power system. The system operates by the transmission of radio frequency energy by conduction over the electric power transmission lines of the system. The system does not include those electric lines which connect the distribution substation to the customer or house wiring.

(u) **Radio frequency (RF) energy.**   Electromagnetic energy at any frequency in the radio spectrum between 9 kHz and 3,000,000 MHz.

(v) **Scanning receiver.**   For the purpose of this Part, this is a receiver that automatically switches among two or more frequencies in the range of 30 to 960 MHz and that is capable of stopping at and receiving a radio signal detected on a frequency. Receivers designed solely for the reception of the broadcast signals under Part 73 of this Chapter, for the reception of NOAA broadcast weather band signals, or for operation as part of a licensed service are not included in this definition.

(w) **Television (TV) broadcast receiver.** A device designed to receive television pictures that are broadcast simultaneously with sound on the television channels authorized under Part 73 of this Chapter.

(x) **Transfer switch.** A device used to alternate between the reception of over-the-air radio frequency signals via connection to an antenna and the reception of radio frequency signals received by any other method, such as from a TV interface device.

(y) **TV interface device.** An unintentional radiator that produces or translates in frequency a radio frequency carrier modulated by a video signal derived from an external or internal signal source and which feeds the modulated radio frequency energy by conduction to the antenna terminals or other non-baseband input connections of a television broadcast receiver. A TV interface device may include a stand-alone RF modulator, or a composite device consisting of an RF modulator, video source, and other components devices. Examples of TV interface devices are videocassette recorders and terminal devices attached to a cable system or used with a Master Antenna (including those used for central distribution video devices in apartment or office buildings).

(z) **Unintentional radiator.** A device that intentionally generates radio frequency energy for use within the device or that sends radio frequency signals by conduction to associated equipment via connecting wiring, but which is not intended to emit RF energy by radiation or induction.

(aa) **Cable ready consumer electronics equipment.** Consumer electronics TV receiving devices, including TV receivers, videocassette recorders and similar devices, that incorporate a tuner capable of receiving television signals and an input terminal for receiving cable television service, and are marketed as "cable ready" or "cable compatible." Such equipment shall comply with the technical standards specified in Section 15.118 of this Chapter.

(bb) **CPU board.**   A circuit board that contains a microprocessor or frequency determining circuitry for the microprocessor, the primary function of which is to execute user-provided programming, but not including:

   (1) a circuit board that contains only a microprocessor intended to operate under the primary control or instruction of a microprocessor external to such a circuit board; or,

   (2) a circuit board that is a dedicated controller for a storage or input/output device.

(cc) **External radio frequency power amplifier.**   A device which is not an integral part of an intentional radiator as manufactured and which, when used in conjunction with an intentional radiator as a signal source, is capable of amplifying that signal.

(dd) **Test Equipment is defined as equipment that is intended primarily for purposes of performing measurements or scientific investigations.**   Such equipment includes, but is not limited to, field strength meters, spectrum analyzers, and modulation monitors.

Section 15.5 General conditions of operation.

(a) Persons operating intentional or unintentional radiators shall not be deemed to have any vested or recognizable right to continued use of any given frequency by virtue of prior registration or certification of equipment, or, for power line carrier systems, on the basis of prior notification of use pursuant to Section 90.63(g) of this Chapter.

(b) Operation of an intentional, unintentional, or incidental radiator is subject to the conditions that no harmful interference is caused and that interference must be accepted that may be caused by the operation of an authorized radio station, by another intentional or unintentional radiator, by industrial, scientific, and medical (ISM) equipment, or by an incidental radiator.

(c) The operator of a radio frequency device shall be required to cease operating the device upon notification by a Commission representative that the device is causing harmful interference. Operation shall not resume until the condition causing the harmful interference has been corrected.

(d) Intentional radiators that produce Class B emissions (damped wave) are prohibited.

Section 15.7 Special temporary authority.

(a) The Commission will, in exceptional situations, consider an individual application for a special temporary authorization to operate an incidental, intentional, or unintentional radiation device not conforming to the provisions of this Part, where it can be shown that the proposed operation would be in the public interest, that it is for a unique type of station or for a type of operation which is incapable of being established as a regular service and that the proposed operation cannot feasibly be conducted under this Part.

(b) No authorization is required in order to perform testing of equipment for determining compliance with these regulations. Except as provided in Subpart I of Part 2 of this Chapter, this provision does not permit the providing of equipment to potential users in order to determine customer acceptance of the product or marketing strategy, nor does this provision permit any type of operation other than a determination of compliance with the regulations. During this testing, the provisions of Sections 15.5 and 15.205 apply.

Section 15.9 Prohibition against eavesdropping.

Except for the operations of law enforcement officers conducted under lawful authority, no person shall use, either directly or indirectly, a device operated pursuant to the provisions of this Part for the purpose of overhearing or recording the private conversations of others unless such use is authorized by all of the parties engaging in the conversation.

Section 15.11 Cross reference.

The provisions of Subparts A, H, I, J, and K of Part 2 apply to intentional and unintentional radiators, in addition to the provisions of this Part. Also, a cable system terminal device and a cable input selector switch shall be subject to the relevant provisions of Part 76 of this Chapter.

Section 15.13 Incidental radiators.

Manufacturers of these devices shall employ good engineering practices to minimize the risk of harmful interference.

Section 15.15 General technical requirements.

(a) An intentional or unintentional radiator shall be constructed in accordance with good engineering design and manufacturing practice. Emanations from the device shall be suppressed as much as practicable, but in no case shall the emanations exceed the levels specified in these rules.

(b) An intentional or unintentional radiator must be constructed such that the adjustments of any control that is readily accessible by or intended to be accessible to the user will not cause operation of the device in violation of the regulations.

(c) Parties responsible for equipment compliance should note that the limits specified in this Part will not prevent harmful interference under all circumstances. Since the operators of Part 15 devices are required to cease operation should harmful interference occur to authorized users of the radio frequency spectrum, the parties responsible for equipment compliance are encouraged to employ the minimum field strength necessary for communications, to provide greater attenuation of unwanted emissions than required by these regulations, and to advise the user as to how to resolve harmful interference problems (for example, see Section 15.105[b]).

Section 15.17 Susceptibility to interference.

(a) Parties responsible for equipment compliance are advised to consider the proximity and the high power of non-government-licensed radio stations, such as broadcast, amateur, land mobile, and non-geostationary mobile satellite feeder link earth stations, and of U.S. government radio stations, which could include high-powered radar systems, when choosing operating frequencies during the design of their equipment so as to reduce the susceptibility for receiving harmful interference. Information on non-government use of the spectrum can be obtained by consulting the Table of Frequency Allocations in § 2.106 of this Chapter.

(b) Information on U. S. government operations can be obtained by contacting: Director, Spectrum Plans and Policy, National Telecommunications and Information Administration, Department of Commerce, Room 4096, Washington, D.C. 20230.

Section 15.19 Labeling requirements.

(a) In addition to the requirements in Part 2 of this Chapter, a device subject to certification or verification shall be labeled as follows:

(1) Receivers associated with the operation of a licensed radio service, e.g. FM broadcast under Part 73, land mobile operation under Part 90, etc. shall bear the following statement in a conspicuous location on the device:

This device complies with Part 15 of the FCC Rules. Operation is subject to the condition that this device does not cause harmful interference.

(2) A stand-alone cable input selector switch shall bear the following statement in a conspicuous location on the device:

This device is verified to comply with Part 15 of the FCC Rules for use with cable television service.

(3) All other devices shall bear the following statement in a conspicuous location on the device:

This device complies with Part 15 of the FCC Rules. Operation is subject to the following two conditions: (1) this device may not cause harmful interference, and (2) this device must accept any interference received, including interference that may cause undesired operation.

(4) Where a device is constructed in two or more sections connected by wires and marketed together, the statement specified under paragraph (a) of this Section is required to be affixed only to the main control unit.

(5) When the device is so small or for such use that it is not practicable to place the statement specified under paragraph (a) of this Section on it, the information required by this paragraph shall be placed in a prominent location in the instruction manual or pamphlet supplied to the user or, alternatively, shall be placed on the container in which the device is marketed. However, the FCC identifier or the unique identifier, as appropriate, must be displayed on the device.

(b) Products subject to authorization under a Declaration of Conformity shall be labeled as follows:

(1) The label shall be located in a conspicuous location on the device and shall contain the unique identification described in Section 2.1074 of this Chapter and the following logo:

(i) If the product is authorized based on testing of the product or system:

Trade Name Model Number

**Tested To Comply
With FCC Standards**

FOR HOME OR OFFICE USE

(ii) If the product is authorized based on assembly using separately authorized components, in accordance with Section 15.101(c)(2) or (c)(3), and the resulting product is not separately tested:

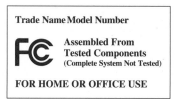

Trade Name Model Number

Assembled From Tested Components (Complete System Not Tested)

FOR HOME OR OFFICE USE

(2) Label text and information should be in a size of type large enough to be readily legible, consistent with the dimensions of the equipment and the label. However, the type size for the text is not required to be larger than eight point.

(3) When the device is so small or for such use that it is not practicable to place the statement specified under paragraph (b)(1) of this Section on it, such as for a CPU board or a plug-in circuit board peripheral device, the text associated with the logo may be placed in a prominent location in the instruction manual or pamphlet supplied to the user. However, the unique identification (trade name and model number) and the logo must be displayed on the device.

(4) The label shall not be a stick-on, paper label. The label on these products shall be permanently affixed to the product and shall be readily visible to the purchaser at the time of purchase, as described in Section 2.925(d) of this Chapter. "Permanently affixed" means that the label is etched, engraved, stamped, silkscreened, indelibly printed, or otherwise permanently marked on a permanently attached part of the equipment or on a nameplate of metal, plastic, or other material fastened to the equipment by welding, riveting, or a permanent adhesive. The label must be designed to last the expected lifetime of the equipment in the environment in which the equipment may be operated and must not be readily detachable.

(c) [reserved]

(d) Consumer electronics TV receiving devices, including TV receivers, videocassette recorders, and similar devices, that incorporate features intended to be used with cable television service, but do not fully comply with the technical standards for cable ready equipment set forth in Section 15.118, shall not be marketed with terminology that describes the device as "cable ready" or "cable compatible," or that otherwise conveys the impression that the device is fully compatible with cable service. Factual statements about the various features of a device that are intended for use with cable service or the quality of such features are acceptable so long as such statements do not imply that the device is fully compatible with cable service. Statements relating to product features are generally acceptable where they are limited to one or more specific features of a device, rather than the device as a whole. This requirement applies to consumer TV receivers, videocassette recorders, and similar devices manufactured or imported for sale in this country on or after October 31, 1994.

Section 15.21 Information to user.

The user manual or instruction manual for an intentional or unintentional radiator shall caution the user that changes or modifications not expressly approved by the party responsible for compliance could void the user's authority to operate the equipment.

Section 15.23 Home-built devices.

(a) Equipment authorization is not required for devices that are not marketed, are not constructed from a kit, and are built in quantities of five or less for personal use.

(b) It is recognized that the individual builder of home-built equipment may not possess the means to perform the measurements for determining compliance with the regulations. In this case, the builder is expected to employ good engineering practices to meet the specified technical standards to the greatest extent practicable. The provisions of Section 15.5 apply to this equipment.

Section 15.25 Kits.

A TV interface device, including a cable system terminal device, which is marketed as a kit shall comply with the following requirements:

(a) All parts necessary for the assembled device to comply with the technical requirements of this Part must be supplied with the kit. No mechanism for adjustment that can cause operation in violation of the requirements of this Part shall be made accessible to the builder.

(b) At least two units of the kit shall be assembled in exact accordance with the instructions supplied with the product to be marketed. If all components required to fully complete the kit (other than those specified in paragraph [a] which are needed for compliance with the technical provisions and must be included with the kit) are not normally furnished with the kit, assembly shall be made using the recommended components. The assembled units shall be certified or authorized under the Declaration of Conformity procedure, as appropriate, pursuant to the requirements of this Part.

  (1) The measurement data required for a TV interface device subject to certification shall be obtained for each of the two units and submitted with an application for certification pursuant to Subpart J of Part 2 of this Chapter.

  (2) The measurement data required for a TV interface device subject to Declaration of Conformity shall be obtained for the units tested and retained on file pursuant to the provisions of Subpart J of Part 2 of this Chapter.

(c) A copy of the exact instructions that will be provided for assembly of the device shall be submitted with an application for certification. Those parts that are not normally furnished shall be detailed in the application for equipment authorization.

(d) In lieu of the label required by Section 15.19, the following label, along with the label bearing the FCC identifier and other information specified in Sections 2.925 and 2.926, shall be included in the kit with instructions to the builder that it shall be attached to the completed kit:

(Name of Grantee)

(FCC Identifier)

This device can be expected to comply with Part 15 of the FCC Rules provided it is assembled in exact accordance with the instructions provided with this kit. Operation is subject to the following conditions: (1) this device may not cause harmful interference, and (2) this device must accept any interference received including interference that may cause undesired operation.

(e) For the purpose of this Section, circuit boards used as repair parts for the replacement of electrically identical defective circuit boards are not considered to be kits.

Section 15.27 Special accessories.

(a) Equipment marketed to a consumer must be capable of complying with the necessary regulations in the configuration in which the equipment is marketed. Where special accessories, such as shielded cables and/or special connectors, are required to enable an unintentional or intentional radiator to comply with the emission limits in this Part, the equipment must be marketed with, i.e., shipped and sold with, those special accessories. However, in lieu of shipping or packaging the special accessories with the unintentional or intentional radiator, the responsible party may employ other methods of ensuring that the special accessories are provided to the consumer, without additional charge, at the time of purchase. Information detailing any alternative method used to supply the special accessories shall be included in the application for a grant of equipment authorization or retained in the verification records, as appropriate. The party responsible for the equipment, as detailed in Section 2.909 of this Chapter, shall ensure that these special accessories are provided with the equipment. The instruction manual for such devices shall include appropriate instructions on the first page of the text concerned with the installation of the device that these special accessories must be used with the device. It is the responsibility of the user to use the needed special accessories supplied with the equipment.

(b) If a device requiring special accessories is installed by or under the supervision of the party marketing the device, it is the responsibility of that party to install the equipment using the special accessories. For equipment requiring professional installation, it is not necessary for the responsible party to market the special accessories with the equipment. However, the need to use the special accessories must be detailed in the instruction manual, and it is the responsibility of the installer to provide and to install the required accessories.

(c) Accessory items that can be readily obtained from multiple retail outlets are not considered to be special accessories and are not required to be marketed with the equipment. The manual included with the equipment must specify what additional components or accessories are required to be used in order to ensure compliance with this Part, and it is the responsibility of the user to provide and use those components and accessories.

(d) The resulting system, including any accessories or components marketed with the equipment, must comply with the regulations.

Section 15.29 Inspection by the Commission.

(a) Any equipment or device subject to the provisions of this Part, together with any certificate, notice of registration, or any technical data required to be kept on file by the operator, supplier, or party responsible for compliance of the device shall be made available for inspection by a Commission representative upon reasonable request.

(b) The owner or operator of a radio frequency device subject to this Part shall promptly furnish to the Commission or its representative such information as may be requested concerning the operation of the radio frequency device.

(c) The party responsible for the compliance of any device subject to this Part shall promptly furnish to the Commission or its representatives such information as may be requested concerning the operation of the device, including a copy of any measurements made for obtaining an equipment authorization or demonstrating compliance with the regulations.

(d) The Commission, from time to time, may request the party responsible for compliance, including an importer, to submit to the FCC Laboratory in Columbia, Maryland, various equipment to determine that the equipment continues to comply with the applicable standards. Shipping costs to the Commission's Laboratory and return shall be borne by the responsible party. Testing by the Commission will be performed using the measurement procedure(s) that was in effect at the time the equipment was authorized or verified.

Section 15.31 Measurement standards.

(a) The following measurement procedures are used by the Commission to determine compliance with the technical requirements in this Part. Except where noted, copies of these procedures are available from the Commission's current duplicating contractor whose name and address are available from the Commission's Consumer Assistance Office at 202-632-7000.

(1) FCC/OET MP-1: FCC Methods of Measurements for Determining Compliance of Radio Control and Security Alarm Devices and Associated Receivers. Note: This procedure may be used only for testing devices for which verification is obtained or for which an application for equipment authorization is filed before June 1, 1995. For compliance testing of these devices after that date, see 15.31(a)(6) of this Section.

(2) FCC/OET MP-2: Measurement of UHF Noise Figures of TV Receivers

(3) FCC/OET MP-3: FCC Methods of Measurements of Output Signal Level, Output Terminal Conducted Spurious Emissions, Transfer Switch Characteristics, and Radio Noise Emissions from TV Interface Devices. Note: This procedure may be used only for testing devices for which verification is obtained or for which an application for equipment authorization is filed before June 1, 1995. For compliance testing of these devices after that date, see 15.31(a)(6) of this Section.

(4) FCC/OET MP-4 (1987): FCC Procedure for Measuring RF Emissions from Computing Devices. Note: This procedure may be used only for testing digital devices for which verification is obtained or for which an application for equipment authorization is filed, before May 1, 1994. For compliance testing of digital devices on or after May 1, 1994, see Section 15.31(a)(6).

(5) FCC/OET MP-9: FCC Procedure for Measuring Cable Television Switch Isolation. Note: This procedure may be used only for testing devices for which verification is obtained or for which an application for equipment authorization is filed before June 1, 1995. For compliance testing of these devices after that date, see 15.31(a)(6) of this Section.

(6) Digital devices authorized by verification, Declaration of Conformity, or for which an application for equipment authorization is filed, on or after May 1, 1994, and intentional and other unintentional radiators for which verification is obtained or for which an application for equipment authorization is filed on or after June 1, 1995, are to be measured for compliance using the following procedure excluding Section 5.7, Section 9, and Section 14: American National Standards Institute (ANSI) C63.4-1992, entitled "Methods of Measurement of Radio-Noise Emissions from Low-Voltage Electrical and Electronic Equipment in the Range of 9 kHz to 40 GHz," published by the Institute of Electrical and Electronics Engineers, Inc. on July 17, 1992, as document number SH15180. This incorporation by reference was approved by the Director of the Federal Register in

accordance with 5 U.S.C. 552(a) and 1 CFR Part 51. The Commission encourages the use of this procedure for testing digital devices, intentional radiators, and other unintentional radiators as soon as practical. Copies of ANSI C63.4-1992 may be obtained from: IEEE Standards Department, 455 Hoes Lane, P.O. Box 1331, Piscataway, NJ 08855-1331, Telephone 1-800-678-4333. Copies of ANSI C63.4-1992 may be inspected during normal business hours at the following locations:

(as stated in the R&O, ANSI C63.4 is not used for measurements below 30 MHz)

(i)   Federal Communications Commission, 2025 M Street, N.W., Office of Engineering and Technology (Room 7317), Washington, D.C. 20554

(ii) Federal Communications Commission Laboratory, 7435 Oakland Mills Road, Columbia, MD 21046

(iii) Office of the Federal Register, 1100 L Street, N.W., Room 8401, Washington, D.C. 20488.

Note: Digital devices tested to show compliance with the provisions of Sections 15.107(e) and 15.109(g) must be tested following the ANSI C63.4 procedure described in paragraph (a)(6).

(b) All parties making compliance measurements on equipment subject to the requirements of this Part are urged to use these measurement procedures. Any party using other procedures should ensure that such other procedures can be relied on to produce measurement results compatible with the FCC measurement procedures. The description of the measurement procedure used in testing the equipment for compliance and a list of the test equipment actually employed shall be made part of an application for certification or included with the data required to be retained by the party responsible for devices authorized pursuant to a Declaration of Conformity or devices subject to Declaration of Conformity (this is how it's now stated in the rules but the duplication is obviously in error) or verification.

(c) For swept frequency equipment, measurements shall be made with the frequency sweep stopped at those frequencies chosen for the measurements to be reported.

(d) Field strength measurements shall be made, to the extent possible, on an open field site. Test sites other than open field sites may be employed if they are properly calibrated so that the measurement results correspond to what would be obtained from an open field site. In the case of equipment for which measurements can be performed only at the installation site, such as perimeter protection systems, carrier current systems, and systems employing a "leaky" coaxial cable as an antenna, measurements for verification, or for obtaining a grant of equipment authorization shall be performed at a minimum of three installations that can be demonstrated to be representative of typical installation sites.

(e) For intentional radiators, measurements of the variation of the input power or the radiated signal level of the fundamental frequency component of the emission, as appropriate, shall be performed with the supply voltage varied between 85% and 115% of the nominal rated supply voltage. For battery-operated equipment, the equipment tests shall be performed using a new battery.

(f) To the extent practicable, the device under test shall be measured at the distance specified in the appropriate rule Section. The distance specified corresponds to the horizontal distance between the measurement antenna and the closest point of the equipment under test, support equipment, or interconnecting cables as determined by the boundary defined by an imaginary straight line periphery describing a simple geometric configuration enclosing the system containing the equipment under test. The equipment under test, support equipment, and any interconnecting cables shall be included within this boundary.

(1) At frequencies at or above 30 MHz, measurements may be performed at a distance other than what is specified provided: measurements are not made in the near field

except where it can be shown that near field measurements are appropriate due to the characteristics of the device; and it can be demonstrated that the signal levels needed to be measured at the distance employed can be detected by the measurement equipment. Measurements shall not be performed at a distance greater than 30 meters unless it can be further demonstrated that measurements at a distance of 30 meters or less are impractical. When performing measurements at a distance other than that specified, the results shall be extrapolated to the specified distance using an extrapolation factor of 20 dB/decade (inverse linear-distance for field strength measurements, inverse-linear-distance-squared for power density measurements).

(2) At frequencies below 30 MHz, measurements may be performed at a distance closer than that specified in the regulations; however, an attempt should be made to avoid making measurements in the near field. Pending the development of an appropriate measurement procedure for measurements performed below 30 MHz, when performing measurements at a closer distance than specified, the results shall be extrapolated to the specified distance by either making measurements at a minimum of two distances on at least one radial to determine the proper extrapolation factor or by using the square of an inverse linear distance extrapolation factor (40 dB/decade).

(3) The applicant for a grant of certification shall specify the extrapolation method used in the application filed with the Commission. For equipment subject to Declaration of Conformity or verification, this information shall be retained with the measurement data.

(4) When measurement distances of 30 meters or less are specified in the regulations, the Commission will test the equipment at the distance specified unless measurement at that distance results in measurements being performed in the near field. When measurement distances of greater than 30 meters are specified in the regulations, the Commission will test the equipment at a closer distance, usually 30

meters, extrapolating the measured field strength to the specified distance using the methods shown in this Section.

(5) Measurements shall be performed at a sufficient number of radials around the equipment under test to determine the radial at which the field strength values of the radiated emissions are maximized. The maximum field strength at the frequency being measured shall be reported in an application for certification.

(g) Equipment under test shall be adjusted, using those controls that are readily accessible to or are intended to be accessible to the consumer, in such a manner as to maximize the level of the emissions. For those devices to which wire leads may be attached by the consumer, tests shall be performed with wire leads attached. The wire leads shall be of the length to be used with the equipment if that length is known. Otherwise, wire leads one meter in length shall be attached to the equipment. Longer wire leads may be employed if necessary to interconnect to associated peripherals.

(h) For a composite system that incorporates devices contained either in a single enclosure or in separate enclosures connected by wire or cable, testing for compliance with the standards in this Part shall be performed with all of the devices in the system functioning. If an intentional radiator incorporates more than one antenna or other radiating source and these radiating sources are designed to emit at the same time, measurements of conducted and radiated emissions shall be performed with all radiating sources that are to be employed emitting. A device which incorporates a carrier current system shall be tested as if the carrier current system were incorporated in a separate device; that is, the device shall be tested for compliance with whatever rules would apply to the device were the carrier current system not incorporated, and the carrier current system shall be tested for compliance with the rules applicable to carrier current systems.

(i) If the device under test provides for the connection of external accessories, including external electrical input signals, the device shall be tested with the accessories attached. The device under test shall be fully exercised with these external accessories. The emission tests shall be performed with the device and accessories configured in a manner that tends to produce maximized emissions within the range of variations that can be expected under normal operating conditions. In the case of multiple accessory external ports, an external accessory shall be connected to one of each type of port. Only one test using peripherals or external accessories that are representative of the devices that will be employed with the equipment under test is required. All possible equipment combinations do not need to be tested. The accessories or peripherals connected to the device being tested shall be unmodified, commercially available equipment.

(j) If the equipment under test consists of a central control unit and an external or internal accessory(ies) (peripheral) and the party verifying the equipment or applying for a grant of equipment authorization manufactures or assembles the central control unit and at least one of the accessory devices that can be used with that control unit, testing of the control unit and/or the accessory(ies) must be performed using the devices manufactured or assembled by that party, in addition to any other needed devices which the party does not manufacture or assemble. If the party verifying the equipment or applying for a grant of equipment authorization does not manufacture or assemble the central control unit and at least one of the accessory devices that can be used with that control unit or the party can demonstrate that the central control unit or accessory(ies) normally would be marketed or used with equipment from a different entity, testing of the central control unit and/or the accessory(ies) must be performed using the specific combination of equipment which is intended to be marketed or used together. Only one test using peripherals or accessories that are representative of the devices that will be employed with the equipment under test is required. All possible equipment combinations are not required to be tested. The accessories or peripherals connected to the device being tested shall be unmodified, commercially available equipment.

(k) A composite system is a system that incorporates different devices contained either in a single enclosure or in separate enclosures connected by wire or cable. If the individual devices in a composite system are subject to different technical standards, each such device must comply with its specific standards. In no event may the measured emissions of the composite system exceed the highest level permitted for an individual component. For digital devices which consist of a combination of Class A and Class B devices, the total combination of which results in a Class A digital device, it is only necessary to demonstrate that the equipment combination complies with the limits for a Class A device. This equipment combination may not be employed for obtaining a grant of equipment authorization or verifying a Class B digital device. However, if the digital device combination consists of a Class B central control unit, e.g., a personal computer, and a Class A internal peripheral(s), it must be demonstrated that the Class B central control unit continues to comply with the limits for a Class B digital device with the Class A internal peripheral(s) installed but not active.

(l) Measurements of radio frequency emissions conducted to the public utility power lines shall be performed using a 50 ohm/50 uH line-impedance stabilization network (LISN). Note: Receivers tested under the transition provisions contained in Section 15.37 may be tested with a 50 ohm/5 uH LISN.

(m) Measurements on intentional radiators or receivers, other than TV broadcast receivers, shall be performed and, if required, reported for each band in which the device can be operated with the device operating at the number of frequencies in each band specified in the following table:

Frequency range over which device operates	Number of frequencies	Location in the range of operation
1 MHz or less	1	Middle
1 to 10 MHz	2	1 near top and 1 near bottom
More than 10 MHz	3	1 near top, 1 near middle, and 1 near bottom

(n) Measurements on TV broadcast receivers shall be performed with the receiver tuned to each VHF frequency and also shall include the following oscillator frequencies: 520, 550, 600, 650, 700, 750, 800, 850, 900, and 931 MHz. If measurements cannot be made on one or more of the latter UHF frequencies because of the presence of signals from licensed radio stations or for other reasons to be detailed in the measurement report, measurements shall be made with the receiver oscillator at a nearby frequency. If the receiver is not capable of receiving channels above 806 MHz, the measurements employing the oscillator frequencies 900 and 931 MHz may be omitted.

(o) The amplitude of spurious emissions from intentional radiators and emissions from unintentional radiators that are attenuated more than 20 dB below the permissible value need not be reported unless specifically required elsewhere in this Part.

(p) In those cases where the provisions in this Section conflict with the measurement procedures in paragraph (a) of this Section and the procedures were implemented after June 23, 1989, the provisions contained in the measurement procedures shall take precedence.

Section 15.32 Test procedures for CPU boards and computer power supplies.

Power supplies and CPU boards used with personal computers and for which separate authorizations are required to be obtained shall be tested as follows:

(a) CPU boards shall be tested as follows:

(1) Testing for radiated emissions shall be performed with the CPU board installed in a typical enclosure, but with the enclosure's cover removed so that the internal circuitry is exposed at the top and on at least two sides. Additional components, including a power supply, peripheral devices, and subassemblies, shall be added, as needed, to result in a complete personal computer system. If the oscillator and the

microprocessor circuits are contained on separate circuit boards, both boards, typical of the combination that would normally be employed, must be used in the test. Testing shall be in accordance with the procedures specified in Section 15.31 of this Part.

(i) Under these test conditions, the system under test shall not exceed the radiated emission limits specified in Section 15.109 of this Part by more than 6 dB. Emissions greater than 6 dB that can be identified and documented to originate from a component(s) other than the CPU board being tested may be dismissed. [Note: When using the test procedure in (a)(1), this test is mandatory and must be passed. Passing the test in (a)(1)(ii) but failing the test here in (a)(1)(i) signifies a non-compliant product. If compliance cannot be shown under (a)(1)(i), see (a)(2) of this Section for an alternative test procedure.]

(ii) Unless the test in paragraph (a)(1)(i) of this Section demonstrates compliance with the limits in Section 15.109 of this part, a second test shall be performed using the same configuration described above, but with the cover installed on the enclosure. Testing shall be in accordance with the procedures specified in Section 15.31 of this part. Under these test conditions, the system under test shall not exceed the radiated emission limits specified in Section 15.109 of this part.

(2) In lieu of the procedure in (a)(1) of this Section, CPU boards may be tested to demonstrate compliance with the limits in Section 15.109 using a specified enclosure with the cover installed. Testing for radiated emissions shall be performed with the CPU board installed in a typical system configuration. Additional components, including a power supply, peripheral devices, and subassemblies, shall be added, as needed, to result in a complete personal computer system. If the oscillator and the microprocessor circuits are contained on separate circuit boards, both boards, typical of the combination that would normally be employed, must be used

in the test. Testing shall be in accordance with the procedures specified in Section 15.31 of this Part. Under this procedure, CPU boards that comply with the limits in Section 15.109 must be marketed together with the specific enclosure used for the test.

(3) The test demonstrating compliance with the AC power line conducted limits specified in Section 15.107 of this Part shall be performed in accordance with the procedures specified in Section 15.31 using a enclosure, peripherals, power supply, and subassemblies that are typical of the type with which the CPU board under test would normally be employed.

(b) The power supply shall be tested installed in an enclosure that is typical of the type within which it would normally be installed. Additional components, including peripheral devices, a CPU board, and subassemblies, shall be added, as needed, to result in a complete personal computer system. Testing shall be in accordance with the procedures specified in Section 15.31 and must demonstrate compliance with all of the standards contained in this part.

Section 15.33 Frequency range of radiated measurements.

(a) Unless otherwise noted in the specific rule Section under which the equipment operates for an intentional radiator the spectrum shall be investigated from the lowest radio frequency signal generated in the device, without going below 9 kHz, up to at least the frequency shown in this paragraph:

(1) If the intentional radiator operates below 10 GHz: to the tenth harmonic of the highest fundamental frequency or to 40 GHz, whichever is lower.

(2) If the intentional radiator operates at or above 10 GHz and below 30 GHz: to the fifth harmonic of the highest fundamental frequency or to 100 GHz, whichever is lower.

(3) If the intentional radiator operates at or above 30 GHz: to the fifth harmonic of the highest fundamental frequency or to

200 GHz, whichever is lower, unless specified otherwise elsewhere in the rules.

(4) If the intentional radiator contains a digital device, regardless of whether this digital device controls the functions of the intentional radiator or the digital device is used for additional control or function purposes other than to enable the operation of the intentional radiator, the frequency range shall be investigated up to the range specified in paragraphs (a)(1)–(a)(3) of this Section or the range applicable to the digital device, as shown in paragraph (b)(1) of this Section, whichever is the higher frequency range of investigation.

(b) For unintentional radiators:

(1) Except as otherwise indicated in paragraphs (b)(2) or (b)(3), for an unintentional radiator, including a digital device, the spectrum shall be investigated from the lowest radio frequency signal generated or used in the device, without going below the lowest frequency for which a radiated emission limit is specified, up to the frequency shown in the following table:

Highest frequency generated or used in the device or on which the device operates or tunes (MHz)	Upper frequency of measurement range (MHz)
Below 1.705	30
1.705–108	1,000
108–500	2,000
500–1,000	5,000
Above 1,000	5th harmonic of the highest frequency or 40 GHz, whichever is lower

(2) A unintentional radiator, excluding a digital device, in which the highest frequency generated in the device, the highest frequency used in the device and the highest frequency on which the device operates or tunes are less than 30 MHz and

which, in accordance with Section 15.109, is required to comply with standards on the level of radiated emissions within the frequency range 9 kHz to 30 MHz, such as a CB receiver or a device designed to conduct its radio frequency emissions via connecting wires or cables, e.g., a carrier current system not intended to radiate, shall be investigated from the lowest radio frequency generated or used in the device, without going below 9 kHz (25 MHz for CB receivers), up to the frequency shown in the following table. If the unintentional radiator contains a digital device, the upper frequency to be investigated shall be that shown in the table below or in the table in paragraph (b)(1) above, as based on both the highest frequency generated and the highest frequency used in the digital device, whichever range is higher.

Highest frequency generated or used in the device or on which the device operates or tunes (MHz)	Upper frequency of measurement range (MHz)
Below 1.705	30
1.705–10	400
10–30	500

(3) Except for a CB receiver, a receiver employing superheterodyne techniques shall be investigated from 30 MHz up to at least the second harmonic of the highest local oscillator frequency generated in the device. If such a receiver is controlled by a digital device, the frequency range shall be investigated up to the higher of the second harmonic of the highest local oscillator frequency generated in the device or the upper frequency of the measurement range specified for the digital device in paragraph (b)(1) of this Section.

(c) The above specified frequency ranges of measurements apply to the measurement of radiated emissions and, in the case of receivers, the measurement to demonstrate compliance with the antenna conduction limits specified in Section 15.111. The frequency range of measurements for AC power line conducted limits is specified in Sections 15.107 and 15.207 and applies to all equipment subject to those regulations. In some cases, depending on the frequency(ies) generated and used by the equipment, only signals conducted onto the AC power lines are required to be measured.

(d) Particular attention should be paid to harmonics and subharmonics of the fundamental frequency as well as to those frequencies removed from the fundamental by multiples of the oscillator frequency. Radiation at the frequencies of multiplier stages should also be checked.

Section 15.35 Measurement detector functions and bandwidths.

The conducted and radiated emission limits shown in this Part are based on the following, unless otherwise specified elsewhere in this Part:

(a) On any frequency or frequencies below or equal to 1,000 MHz, the limits shown are based on measuring equipment employing a CISPR quasi-peak detector function and related measurement bandwidths, unless otherwise specified. The specifications for the measuring instrument using the CISPR quasi-peak detector can be found in Publication 16 of the International Special Committee on Radio Interference (CISPR) of the International Electrotechnical Commission. As an alternative to CISPR quasi-peak measurements, the responsible party, at its option, may demonstrate compliance with the emission limits using measuring equipment employing a peak detector function, properly adjusted for such factors as pulse desensitization, as long as the same bandwidths as indicated for CISPR quasi-peak measurements are employed.

Note: For pulse-modulated devices with a pulse-repetition frequency of 20 Hz or less and for which CISPR quasi-peak measurements are specified, compliance with the regulations shall be demonstrated using measuring equipment employing a peak detector function, properly adjusted for such factors as pulse desensitization, using the same measurement bandwidths that are indicated for CISPR quasi-peak measurements.

(b) On any frequency of frequencies above 1,000 MHz, the radiated limits shown are based upon the use of measurement instrumentation employing an average detector function. When average radiated emission measurements are specified in the regulations, including emission measurements below 1,000 MHz, there is also a limit on the radio frequency emissions, as measured using instrumentation with a peak detector function, corresponding to 20 dB above the maximum permitted average limit for the frequency being investigated unless a different peak emission limit is otherwise specified in the rules, e.g., see Section 15.255. Unless otherwise specified, measurements above 1,000 MHz shall be performed using a minimum resolution bandwidth of 1 MHz. Measurement of AC power line conducted emissions are performed using a CISPR quasi-peak detector, even for devices for which average radiated emission measurements are specified.

(c) Unless otherwise specified, e.g. Section 15.255(b), when the radiated emission limits are expressed in terms of the average value of the emission and pulsed operation is employed, the measurement field strength shall be determined by averaging over one complete pulse train, including blanking intervals, as long as the pulse train does not exceed 0.1 seconds. As an alternative (provided the transmitter operates for longer than 0.1 seconds) or, in cases where the pulse train exceeds 0.1 seconds, the measured field strength shall be determined from the average absolute voltage during a 0.1 second interval during which the field strength is at its maximum value. The exact method of calculating the average field strength shall be submitted with any application for certification or shall be retained in the measurement data file for equipment subject to notification or verification.

Section 15.37 Transition provisions for compliance with the rules.

Equipment may be authorized, manufactured, and imported under the rules in effect prior to June 23, 1989, in accordance with the following schedules:

(a) **For all intentional and unintentional radiators, except for receivers:** Radio frequency equipment verified by the responsible party or for which an application for a grant of equipment authorization is submitted to the Commission on or after June 23, 1992, shall comply with the regulations specified in this Part. Radio frequency equipment that is manufactured or imported on or after June 23, 1994, shall comply with the regulations specified in this Part.

(b) **For receivers:** Receivers subject to the regulations in this Part that are manufactured or imported on or after June 23, 1999, shall comply with the regulations specified in this Part. However, if a receiver is associated with a transmitter that could not have been authorized under the regulations in effect prior to June 23, 1989, e.g., a transmitter operating under the provisions of Sections 15.209 or 15.249 (below 960 MHz), the transition provisions in this Section do not apply. Such receivers must comply with the regulations in this Part. In addition, receivers are subject to the provisions in paragraph (f) of this Section.

(c) There are no restrictions on the operation or marketing of equipment complying with the regulations in effect prior to June 23, 1989.

(d) Prior to May 25, 1991, no person shall import, market, or operate intentional radiators within the band 902–905 MHz under the provisions of Section 15.249. Until that date, the Commission will not issue a grant of equipment authorization for equipment operating under Section 15.249 if the equipment is designed to permit operation within the band 902–905 MHz.

(e) **For cordless telephones:** The manufacture and importation of cordless telephones not complying with Section 15.214(d) of this Part shall cease on or before September 11, 1991. These provisions will not apply to cordless telephones which are repaired or refurbished, or re-imported after repair or refurbishment. Applications for a grant of equipment authorization of cordless telephones not complying with Section 15.214(d) of this Part will not be accepted by the Commission after May 10, 1991. Cordless telephones that have previously received equipment authorization and that, without modification, already comply with the requirements of Section 15.214(d) of this Part need not be reauthorized.

(f) The manufacture or importation of scanning receivers and frequency converters designed or marketed for use with scanning receivers that do not comply with the provisions of § 15.121(a)(1) shall cease on or before April 26, 1994. Effective April 26, 1993, the Commission will not grant equipment authorization for receivers that do not comply with the provisions of § 15.121(a)(1). This paragraph does not prohibit the sale or use of authorized receivers manufactured in the United States, or imported into the United States, prior to April 26, 1994.

(g) **For CPU boards and power supplies designed to be used with personal computers:** The manufacture and importation of these products shall cease on or before September 17, 1997, unless these products have been authorized under a Declaration of Conformity or a grant of certification, demonstrating compliance with all of the provisions in this Part. Limited provisions, as detailed in Section 15.101(d) of this Chapter, are provided to permit the importation and manufacture of these products subsequent to this date where the CPU boards and/or power supplies are marketed only to personal computer equipment manufacturers.

(h) The manufacture or importation of scanning receivers, and frequency converters designed or marketed for use with scanning receivers, that do not comply with the provisions of § 15.121 shall cease on or before October 25, 1999. Effective

July 26, 1999, the Commission will not grant equipment authorization for receivers that do not comply with the provisions of § 15.121. This paragraph does not prohibit the sale or use of authorized receivers manufactured in the United States, or imported into the United States, prior to October 25, 1999.

(i) Effective October 16, 2002, an equipment approval may no longer be obtained for medical telemetry equipment operating under the provisions of § 15.241 or § 15.242. The requirements for obtaining an approval for medical telemetry equipment after this date are found in Subpart H of Part 95 of this Chapter.

Subpart C—Intentional Radiators

Section 15.201 Equipment authorization requirement.

(a) Intentional radiators operated as carrier current systems and devices operated under the provisions of Sections 15.211, 15.213, and 15.221 shall be verified pursuant to the procedures in Subpart J of Part 2 of this Chapter prior to marketing.

(b) Except as otherwise exempted in paragraph (c) of this Section and in Section 15.23 of this Part, all intentional radiators operating under the provisions of this Part shall be certificated by the Commission pursuant to the procedures in Subpart J of Part 2 of this Chapter prior to marketing.

(c) For devices such as perimeter protection, systems which, in accordance with Section 15.31(d), are required to be measured at the installation site, each application for certification must be accompanied by a statement indicating that the system has been tested at three installations and found to comply at each installation. Until such time as certification is granted, a given installation of a system that was measured for the submission for certification will be considered to be in compliance with the provisions of this Chapter, including the marketing regulations in Subpart I of Part 2, if tests at that installation show the system to be in compliance with the relevant technical requirements. Similarly, where measurements must be performed on site for equipment subject to verification, a given installation that has been verified to demonstrate compliance with the applicable standards will be considered to be in compliance with the provisions of this Chapter, including the marketing regulations in Subpart I of Part 2.

(d) For perimeter protection systems operating in the frequency bands allocated to television broadcast stations operating under Part 73 of this Chapter, the holder of the grant of certification must test each installation prior to initiation of normal operation to verify compliance with the technical standards and must maintain a list of all installations and records of measurements. For perimeter protection systems operating outside of the frequency bands allocated to television broadcast stations, upon receipt of a grant of certification, further testing of the same or similar type of system or installation is not required.

Section 15.203 Antenna requirement.

An intentional radiator shall be designed to ensure that no antenna other than that furnished by the responsible party shall be used with the device. The use of a permanently attached antenna or of an antenna that uses a unique coupling to the intentional radiator shall be considered sufficient to comply with the provisions of this Section. The manufacturer may design the unit so that a broken antenna can be replaced by the user, but the use of a standard antenna jack or electrical connector is prohibited. This requirement does not apply to carrier current devices or to devices operated under the provisions of Sections 15.211, 15.213, 15.217, 15.219, or 15.221. Further, this requirement does not apply to intentional radiators that must be professionally installed, such as perimeter protection systems and some field disturbance sensors, or to other intentional radiators which, in accordance with Section 15.31(d), must be measured at the installation site. However, the installer shall be responsible for ensuring that the proper antenna is employed so that the limits in this Part are not exceeded.

Section 15.204 External radio frequency power amplifiers and antenna modifications.

(a) Except as otherwise described in paragraph (b) of this Section, no person shall use, manufacture, sell or lease, offer for sale or lease (including advertising for sale or lease), or import, ship, or distribute for the purpose of selling or leasing any external radio frequency power amplifier or amplifier kit intended for use with a Part 15 intentional radiator.

(b) A transmission system consisting of an intentional radiator, an external radio frequency power amplifier, and an antenna, may be authorized, marketed, and used under this part. However, when a transmission system is authorized as a system, it must always be marketed as a complete system and must always be used in the configuration in which it was authorized. An external radio frequency power amplifier shall be marketed only in the system configuration with which the amplifier is authorized and shall not be marketed as a separate product.

(c) Only the antenna with which an intentional radiator is authorized may be used with the intentional radiator.

Section 15.205 Restricted bands of operation.

(a) Except as shown in paragraph (d) of this Section, only spurious emissions are permitted in any of the frequency bands listed below:

MHz	MHz	MHz	GHz
0.090–0.110	16.42–16.423	399.9–410	4.5–5.15
[1]0.495–0.505	16.69475–16.69525	608–614	5.35–5.46
2.1735–2.1905	16.80425–16.80475	960–1,240	7.25–7.75

MHz	MHz	MHz	GHz
4.125–4.128	25.5–25.67	1,300–1,427	8.025–8.5
4.17725–4.17775	37.5–38.25	1,435–1,626.5	9.0–9.2
4.20725–4.20775	73–74.6	1,645.5–1,646.5	9.3–9.5
6.215–6.218	74.8–75.2	1,660–1,710	10.6–12.7
6.26775–6.26825	108–121.94	1,718.8–1,722.2	13.25–13.4
6.31175–6.31225	123–138	2,200–2,300	14.47–14.5
8.291–8.294	149.9–150.05	2,310–2,390	15.35–16.2
8.362–8.366	156.52475–156.52525	2,483.5–2,500	17.7–21.4
8.37625–8.38675	156.7–156.9	2,655–2,900	22.01–23.12
8.41425–8.41475	162.0125–167.17	3,260–3,267	23.6–24.0
12.29–12.293	167.72–173.2	3,332–3,339	31.2–31.8
12.51975–12.52025	240–285	3,345.8–3,358	36.43–36.5
12.57675–12.57725	322–335.4	3,600–4,400	(2)
13.36–13.41			

[1]Until February 1, 1999, this restricted band shall be 0.490–0.510 MHz.
[2]Above 38.6

(b) Except as provided in paragraphs (d) and (e), the field strength of emissions appearing within these frequency bands shall not exceed the limits shown in Section 15.209. At frequencies equal to or less than 1,000 MHz, compliance with the limits in Section 15.209 shall be demonstrated using measurement instrumentation employing a CISPR quasi-peak detector. Above 1,000 MHz, compliance with the emission limits in Section 15.209 shall be demonstrated based on the average value of the measured emissions. The provisions in Section 15.35 apply to these measurements.

(c) Except as provided in paragraphs (d) and (e), regardless of the field strength limits specified elsewhere in this Subpart, the provisions of this Section apply to emissions from any intentional radiator.

(d) The following devices are exempt from the requirements of this Section:

    (1) Swept frequency field disturbance sensors operating between 1.705 and 37 MHz provided their emissions only sweep through the bands listed in paragraph (a), the sweep is never stopped with the fundamental emission within the bands listed in paragraph (a), and the fundamental emission is outside of the bands listed in paragraph (a) more than 99% of the time the device is actively transmitting, without compensation for duty cycle.

    (2) Transmitters used to detect buried electronic markers at 101.4 kHz that are employed by telephone companies.

    (3) Cable locating equipment operated pursuant to Section 15.213.

    (4) Any equipment operated under the provisions of Section 15.253 or Section 15.255 of this Part.

    (5) Biomedical telemetry devices operating under the provisions of Section 15.242 of this Part are not subject to the restricted band 608–614 MHz, but are subject to compliance within the other restricted bands.

(e) Harmonic emissions appearing in the restricted bands above 17.7 GHz from field disturbance sensors operating under the provisions of Section 15.245 shall not exceed the limits specified in Section 15.245(b).

Section 15.207 Conducted limits.

(a) For an intentional radiator that is designed to be connected to the public utility (AC) power line, the radio frequency voltage that is conducted back onto the AC power line on any frequency or frequencies within the band 450 kHz to 30 MHz shall not exceed 250 microvolts. Compliance with this provision shall be based on the measurement of the radio frequency voltage between each power line and ground at the power terminals.

(b) The following option may be employed if the conducted emissions exceed the limits in paragraph (a) of this Section when measured using instrumentation employing a quasi-peak detector function: if the level of the emission measured using the quasi-peak instrumentation is 6 dB, or more, higher than the level of the same emission measured with instrumentation having an average detector and a 9 kHz minimum bandwidth, that emission is considered broadband and the level obtained with the quasi-peak detector may be reduced by 13 dB for comparison to the limits. When employing this option, the following conditions shall be observed:

   (1) The measuring instrumentation with the average detector shall employ a linear IF amplifier.

   (2) Care must be taken not to exceed the dynamic range of the measuring instrument when measuring an emission with a low duty cycle.

   (3) The test report required for verification or for an application for a grant of equipment authorization shall contain all details supporting the use of this option.

(c) The shown limit in paragraph (a) of this Section shall not apply to carrier current systems operating as intentional radiators on frequencies below 30 MHz. In lieu thereof, these carrier current systems shall be subject to the following standards:

   (1) For carrier current systems containing their fundamental emission within the frequency band 535–1,705 kHz and intended to be received using a standard AM broadcast receiver: no limit on conducted emissions.

   (2) For all other carrier current systems: 1,000 uV within the frequency band 535–1,705 kHz.

   (3) Carrier current systems operating below 30 MHz are also subject to the radiated emission limits in Section 15.205 and Section 15.209, 15.221, 15.223, 15.225, or 15.227, as appropriate.

(d) Measurements to demonstrate compliance with the conducted limits are not required for devices which only employ battery power for operation and which do not operate from the AC power lines or contain provisions for operation while connected to the AC power lines. Devices that include, or make provision for, the use of battery chargers which permit operating while charging, AC adaptors, or battery eliminators or that connect to the AC power lines indirectly, obtaining their power through another device which is connected to the AC power lines, shall be tested to demonstrate compliance with the conducted limits.

Section 15.209 Radiated emission limits, general requirements.

(a) Except as provided elsewhere in this Subpart, the emissions from an intentional radiator shall not exceed the field strength levels specified in the following table:

Frequency (MHz)	Field Strength (microvolts/meter)	Measurement Distance (meters)
0.009–0.490	2,400/F(kHz)	300
0.490–1.705	24,000/F(kHz)	30
1.705–30.0	30	30
30–88	100**	3
88–216	150**	3
216–960	200**	3
Above 960	500	3

**Except as provided in paragraph (g), fundamental emissions from intentional radiators operating under this Section shall not be located in the frequency bands 54–72 MHz, 76–88 MHz, 174–216 MHz, or 470–806 MHz. However, operation within these frequency bands is permitted under other Sections of this Part, e.g., Sections 15.231 and 15.241.

(b) In the emission table above, the tighter limit applies at the band edges.

(c) The level of any unwanted emissions from an intentional radiator operating under these general provisions shall not exceed the level of the fundamental emission. For intentional radiators which operate under the provisions of other Sections within this Part and which are required to reduce their unwanted emissions to the limits specified in this table, the limits in this table are based on the frequency of the unwanted emission and not the fundamental frequency. However, the level of any unwanted emissions shall not exceed the level of the fundamental frequency.

(d) The emission limits shown in the above table are based on measurements employing a CISPR quasi-peak detector except for the frequency bands 9–90 kHz, 110–490 kHz, and above 1,000 MHz. Radiated emission limits in these three bands are based on measurements employing an average detector.

(e) The provisions in Sections 15.31, 15.33, and 15.35 for measuring emissions at distances other than the distances specified in the above table, determining the frequency range over which radiated emissions are to be measured and limiting peak emissions apply to all devices operated under this Part.

(f) In accordance with Section 15.33(a), in some cases, the emissions from an intentional radiator must be measured to beyond the tenth harmonic of the highest fundamental frequency designed to be emitted by the intentional radiator because of the incorporation of a digital device. If measurements above the tenth harmonic are so required, the radiated emissions above the tenth harmonic shall comply with the general radiated emission limits applicable to the incorporated digital device, as shown in Section 15.109 and as based on the frequency of the emission being measured, or, except for emissions contained in the restricted frequency bands shown in Section 15.205, the limit on spurious emissions specified for the intentional radiator, whichever is the higher limit. Emissions which must be measured above the tenth harmonic of the highest fundamental frequency designed to be emitted by the intentional radiator and which fall within the restricted bands shall comply with the general radiated emission limits in Section 15.109 that are applicable to the incorporated digital device.

(g) Perimeter protection systems may operate in the 54–72 MHz and 76–88 MHz bands under the provisions of this Section. The use of such perimeter protection systems is limited to industrial, business, and commercial applications.

Section 15.211 Tunnel radio systems.

An intentional radiator utilized as part of a tunnel radio system may operate on any frequency provided it meets all of the following conditions:

(a) Operation of a tunnel radio system (intentional radiator and all connecting wires) shall be contained solely within a tunnel, mine, or other structure that provides attenuation to the radiated signal due to the presence of naturally surrounding earth and/or water.

(b) Any intentional or unintentional radiator external to the tunnel, mine, or other structure, as described in paragraph (a), shall be subject to the other applicable regulations contained within this part.

(c) The total electromagnetic field from a tunnel radio system on any frequency or frequencies appearing outside of the tunnel, mine or other structure described in paragraph (a) shall not exceed the limits shown in Section 15.209 when measured at the specified distance from the surrounding structure, including openings. Particular attention shall be paid to the emissions from any opening in the structure to the outside environment. When measurements are made from the openings, the distances shown in Section 15.209 refer to the distance from the plane of reference that fits the entire perimeter of each above ground opening.

(d) The conducted limits in Section 15.207 apply to the radio frequency voltage on the public utility power lines outside of the tunnel.

Section 15.213 Cable locating equipment.

An intentional radiator used as cable locating equipment, as defined in Section 15.3(d), may be operated on any frequency within the band 9–490 kHz, subject to the following limits: within the frequency band 9 kHz, up to, but not including, 45 kHz, the peak output power from the cable locating equipment shall not exceed 10 watts; and, within the frequency band 45 kHz to 490 kHz, the peak output power from the cable locating equipment shall not exceed one watt. If provisions are made for connection of the cable locating equipment to the AC power lines, the conducted limits in Section 15.207 also apply to this equipment.

Section 15.214 Cordless telephones.

(a) For equipment authorization, a single application form, FCC Form 731, may be filed for a cordless telephone system, provided the application clearly identifies and provides data for all parts of the system to show compliance with the applicable technical requirements. When a single application form is submitted, both the base station and the portable handset must carry the same FCC identifier. The application shall include a fee for certification of each type of transmitter and for certification, if appropriate, for each type of receiver included in the system.

(b) A cordless telephone that is intended to be connected to the public switched telephone network shall also comply with the applicable regulations in Part 68 of this Chapter. A separate application for registration under Part 68 of this Chapter is required.

(c) The label required under Subpart A of this Part shall also contain the following statement:

"Privacy of communications may not be ensured when using this phone."

(d) Cordless telephones shall incorporate circuitry that makes use of a digital security code to provide protection against unintentional access to the public switched telephone network by the base unit and unintentional ringing by the handset. These functions shall operate such that each access of the telephone network or ringing of the handset is preceded by the transmission of a code word. Access to the telephone network shall occur only if the code transmitted by the handset matches the code set in the base unit. Similarly, ringing of the handset shall occur only if the code transmitted by the base unit matches the code set in the handset. The security code required by this Section may also be employed to perform other communications functions, such as providing telephone billing information. This security code system is to operate in accordance with the following provisions.

(1) There must be provision for at least 256 possible discrete digital codes. Factory-set codes must be continuously varied over at least 256 possible codes as each telephone is manufactured. The codes may be varied either randomly, sequentially, or using another systematic procedure.

(2) Manufacturers must use one of the following approaches for facilitating variation in the geographic distribution of individual security codes:

　(i) Provide a means for the user to readily select from among at least 256 possible discrete digital codes. The cordless telephone shall be either in a non-operable mode after manufacture until the user selects a security code or the manufacturer must continuously vary the initial security code as each telephone is produced.

　(ii) Provide a fixed code that is continuously varied among at least 256 discrete digital codes as each telephone is manufactured.

　(iii) Provide a means for the cordless telephone to automatically select a different code from among at least 256 possible discrete digital codes each time it is activated.

　(iv) It is permissible to provide combinations of fixed, automatic, and user-selectable coding provided the above criteria are met.

(3) A statement of the means and procedures used to achieve the required protection shall be provided in any application for equipment authorization of a cordless telephone.

Section 15.247 Operation within the bands 902–928 MHz, 2,400–2,483.5 MHz, and 5,725–5,850 MHz.

(a) Operation under the provisions of this Section is limited to frequency hopping and direct sequence spread spectrum intentional radiators that comply with the following provisions:

(1) Frequency hopping systems shall have hopping channel carrier frequencies separated by a minimum of 25 kHz or the 20 dB bandwidth of the hopping channel, whichever is greater. The system shall hop to channel frequencies that are selected at the system hopping rate from a pseudorandomly ordered list of hopping frequencies. Each frequency must be used equally on the average by each transmitter. The system receivers shall have input bandwidths that match the hopping channel bandwidths of their corresponding transmitters and shall shift frequencies in synchronization with the transmitted signals.

(i) For frequency hopping systems operating in the 902–928 MHz band: if the 20 dB bandwidth of the hopping channel is less than 250 kHz, the system shall use at least 50 hopping frequencies and the average time of occupancy on any frequency shall not be greater than 0.4 seconds within a 20 second period; if the 20 dB bandwidth of the hopping channel is 250 kHz or greater, the system shall use at least 25 hopping frequencies and the average time of occupancy on any frequency shall not be greater than 0.4 seconds within a 10 second period. The maximum allowed 20 dB bandwidth of the hopping channel is 500 kHz.

(ii) Frequency hopping systems operating in the 2,400–2,483.5 MHz and 5,725–5,850 MHz bands shall use at least 75 hopping frequencies. The maximum 20 dB bandwidth of the hopping channel is 1 MHz. The average time of occupancy on any frequency shall not be greater than 0.4 seconds within a 30 second period.

(iii) Frequency hopping systems in the 2,400–2,483.5 MHz band may utilize hopping channels whose 20 dB bandwidth is greater than 1 MHz provided the systems use at least 15 non-overlapping channels. The total span of hopping channels shall be at least 75 MHz. The time of occupancy on any one channel shall be no greater than 0.4 seconds within the time period required to hop through all channels. (Note that each of the hopping channels must be used equally on the average.)

(2) For direct sequence systems, the minimum 6 dB bandwidth shall be at least 500 kHz.

(b) The maximum peak output power of the intentional radiator shall not exceed the following:

(1) For frequency hopping systems operating in the 2,400–2,483.5 MHz band employing at least 75 hopping channels, all frequency hopping systems in the 5,725–5,850 MHz band and all direct sequence systems: 1 watt. For all other frequency hopping systems in the 2,400–2,483.5 MHz band: 0.125 watts.

(2) For frequency hopping systems operating in the 902–928 MHz band: 1 watt for systems employing at least 50 hopping channels and 0.25 watts for systems employing less than 50 hopping channels, but at least 25 hopping channels, as permitted under paragraph (a)(1)(i) of this Section.

(3) Except as shown below, if transmitting antennas of directional gain greater than 6 dBi are used the peak output power from the intentional radiator shall be reduced below the above stated values by the amount in dB that the directional gain of the antenna exceeds 6 dBi.

(i) Systems operating in the 2,400–2,483.5 MHz band that are used exclusively for fixed, point-to-point operations may employ transmitting antennas with directional gain greater than 6 dBi provided the maximum peak output power of the intentional radiator is reduced by 1 dB for every 3 dB that the directional gain of the antenna exceeds 6 dBi.

(ii) Systems operating in the 5,725–5,850 MHz band that are used exclusively for fixed, point-to-point operations may employ transmitting antennas with directional gain greater than 6 dBi without any corresponding reduction in transmitter peak output power.

(iii) Fixed, point-to-point operation, as used in paragraphs (b)(3)(i) and (b)(3)(ii) of this Section, excludes the use of point-to-multipoint systems, omnidirectional applications, and multiple co-located intentional radiators transmitting the same information. The operator of the spread spectrum intentional radiator or, if the equipment is professionally installed, the installer is responsible for ensuring that the system is used exclusively for fixed, point-to-point operations. The instruction manual furnished with the intentional radiator shall contain language in the installation instructions informing the operator and the installer of this responsibility.

(4) Systems operating under the provisions of this Section shall be operated in a manner that ensures that the public is not exposed to radio frequency energy levels in excess of the Commission's guidelines. See § 1.1307(b)(1) of this Chapter.

(c) In any 100 kHz bandwidth outside the frequency band in which the spread spectrum intentional radiator is operating, the radio frequency power that is produced by the intentional radiator shall be at least 20 dB below that in the 100 kHz bandwidth within the band that contains the highest level of the desired power, based on either an RF conducted or a radiated measurement. Attenuation below the general limits specified in Section 15.209(a) is not required. In addition, radiated emissions that fall in the restricted bands, as defined in Section 15.205(a), must also comply with the radiated emission limits specified in Section 15.209(a) (see Section 15.205[c]).

(d) For direct sequence systems, the peak power spectral density conducted from the intentional radiator to the antenna shall not be greater than 8 dBm in any 3 kHz band during any time interval of continuous transmission.

(e) The processing gain of a direct sequence system shall be at least 10 dB. The processing gain represents the improvement to the received signal-to-noise ratio, after filtering to the information bandwidth, from the spreading/despreading function. The processing gain may be determined using one of the following methods:

(1) As measured at the demodulated output of the receiver: The ratio in dB of the signal-to-noise ratio with the system spreading code turned off to the signal-to-noise ratio with the system spreading code turned on.

(2) As measured using the CW jamming margin method: a signal generator is stepped in 50 kHz increments across the passband of the system, recording at each point the generator level required to produce the recommended Bit Error Rate (BER). This level is the jammer level. The output power of the intentional radiator is measured at the same point. The jammer to signal ratio (J/S) is then calculated, discarding the worst 20% of the J/S data points. The lowest remaining J/S ratio is used to calculate the processing gain, as follows: Gp = (S/N)o + Mj + Lsys, where Gp = processing gain of the system, (S/N)o = signal-to-noise ratio required for the chosen BER, Mj = J/S ratio, and Lsys = system losses. Note that total losses in a system, including intentional radiator and receiver, should be assumed to be no more than 2 dB.

(f) Hybrid systems that employ a combination of both direct sequence and frequency hopping modulation techniques shall achieve a processing gain of at least 17 dB from the combined techniques. The frequency hopping operation of the hybrid system, with the direct sequence operation turned off, shall have an average time of occupancy on any frequency not to exceed 0.4 seconds within a time period in seconds equal to the number of hopping frequencies employed multiplied by 0.4. The direct sequence operation of the hybrid system, with the frequency hopping turned off, shall comply with the power density requirements of paragraph (d) of this Section.

(g) Frequency hopping spread spectrum systems are not required to employ all available hopping channels during each transmission. However, the system, consisting of both the transmitter and the receiver, must be designed to comply with all of the regulations in this Section should the transmitter be presented with a continuous data (or information) stream. In addition, a system employing short transmission bursts must comply with the definition of a frequency hopping system and must distribute its transmissions over the minimum number of hopping channels specified in this Section.

(h) The incorporation of intelligence within a frequency hopping spread spectrum system that permits the system to recognize other users within the spectrum band so that it individually and independently chooses and adapts its hopsets to avoid hopping on occupied channels is permitted. The coordination of frequency hopping systems in any other manner for the express purpose of avoiding the simultaneous occupancy of individual hopping frequencies by multiple transmitters is not permitted.

Note: Spread spectrum systems are sharing these bands on a noninterference basis with systems supporting critical government requirements that have been allocated the usage of these bands, secondary only to ISM equipment operated under the provisions of Part 18 of this Chapter. Many of these government systems are airborne radiolocation systems that emit a high EIRP that can cause interference to other users. Also, investigations of the effect of spread spectrum interference to U. S. government operations in the 902–928 MHz band may require a future decrease in the power limits allowed for spread spectrum operation.

Subpart E—Unlicensed National Information Infrastructure Devices

Section 15.401 Scope.

This Subpart sets out the regulations for unlicensed National Information Infrastructure (U-NII) devices operating in the 5.15–5.35 GHz and 5.725–5.825 GHz bands.

Section 15.403 Definitions.

(a) **Average symbol envelope power.** The average symbol envelope power is the average, taken over all symbols in the signaling alphabet, of the envelope power for each symbol.

(b) **Digital modulation.** The process by which the characteristics of a carrier wave are varied among a set of predetermined discrete values in accordance with a digital modulating function as specified in document ANSI C63.17-1998.

(c) **Emission bandwidth.** For purposes of this Subpart, the emission bandwidth shall be determined by measuring the width of the signal between two points, one below the carrier center frequency and one above the carrier center frequency, that are 26 dB down relative to the maximum level of the modulated carrier. Determination of the emissions bandwidth is based on the use of measurement instrumentation employing a peak detector function with an instrument resolution's bandwidth approximately equal to 1.0 percent of the emission bandwidth of the device under measurement.

(d) **Peak power spectral density.** The peak power spectral density is the maximum power spectral density, within the specified measurement bandwidth, within the U-NII device operating band.

(e) **Peak transmit power.** The maximum transmit power as measured over an interval of time of at most 30/B or the transmission pulse duration of the device, whichever is less, under all conditions of modulation.

(f) **Power spectral density.** The power spectral density is the total energy output per unit bandwidth from a pulse or sequence of pulses for which the transmit power is at its peak or maximum level, divided by the total duration of the pulses. This total time does not include the time between pulses during which the transmit power is off or below its maximum level.

(g) **Pulse.** A pulse is a continuous transmission of a sequence of modulation symbols, during which the average symbol envelope power is constant.

(h) **Transmit power.** The total energy transmitted over a time interval of at most 30/B (where B is the 26 dB emission bandwidth of the signal in hertz) or the duration of the transmission pulse, whichever is less, divided by the interval duration.

(i) **U-NII devices.** Intentional radiators operating in the frequency bands 5.15–5.35 GHz and 5.725–5.825 GHz that use wideband digital modulation techniques and provide a wide array of high data rate mobile and fixed communications for individuals, businesses, and institutions.

Section 15.405 Cross reference.

(a) The provisions of Subparts A, B, and C of this Part apply to unlicensed U-NII devices, except where specific provisions are contained in Subpart E. Manufacturers should note that this includes the provisions of Sections 15.203 and 15.205.

(b) The requirements of Subpart E apply only to the radio transmitter contained in the U-NII device. Other aspects of the operation of a U-NII device may be subject to requirements contained elsewhere in this Chapter. In particular, a U-NII device that includes digital circuitry not directly associated with the radio transmitter also is subject to the requirements for unintentional radiators in Subpart B.

Section 15.407 General technical requirements.

(a) Power limits:

    (1) For the band 5.15–5.25 GHz, the peak transmit power over the frequency band of operation shall not exceed the lesser of 50 mW or 4 dBm + 10log B, where B is the 26-dB emission bandwidth in MHz. In addition, the peak power spectral density shall not exceed 4 dBm in any 1-MHz band. If transmitting antennas of directional gain greater than 6 dBi are used, both the peak transmit power and the peak power spectral density shall be reduced by the amount in dB that the directional gain of the antenna exceeds 6 dBi.

    (2) For the band 5.25–5.35 GHz, the peak transmit power over the frequency band of operation shall not exceed the lesser of 250 mW or 11 dBm + 10log B, where B is the 26-dB emission bandwidth in MHz. In addition, the peak power spectral density shall not exceed 11 dBm in any 1-MHz band. If transmitting antennas of directional gain greater than 6 dBi are used, both the peak transmit power and the peak power spectral density shall be reduced by the amount in dB that the directional gain of the antenna exceeds 6 dBi.

    (3) For the band 5.725–5.825 GHz, the peak transmit power over the frequency band of operation shall not exceed the lesser of 1 W or 17 dBm + 10log B, where B is the 26-dB emission bandwidth in MHz. In addition, the peak power spectral density shall not exceed 17 dBm in any 1-MHz band. If transmitting antennas of directional gain greater than 6 dBi are used, both the peak transmit power and the peak power spectral density shall be reduced by the amount in dB that the directional gain of the antenna exceeds 6 dBi. However, fixed point-to-point U-NII devices operating in this band may employ transmitting antennas with directional gain up to 23

dBi without any corresponding reduction in the transmitter peak output power or peak power spectral density. For fixed, point-to-point U-NII transmitters that employ a directional antenna gain greater than 23 dBi, a 1 dB reduction in peak transmitter power and peak power spectral density for each 1 dB of antenna gain in excess of 23 dBi would be required. Fixed, point-to-point operations exclude the use of point-to-multipoint systems, omnidirectional applications, and multiple collocated transmitters transmitting the same information. The operator of the U-NII device, or if the equipment is professionally installed, the installer, is responsible for ensuring that systems employing high gain directional antennas are used exclusively for fixed, point-to-point operations.

Note: The Commission strongly recommends that parties employing U-NII devices to provide critical communications services should determine if there are any nearby government radar systems that could affect their operation.

(4) The peak transmit power must be measured over any interval of continuous transmission using instrumentation calibrated in terms of an rms-equivalent voltage. The measurement results shall be properly adjusted for any instrument limitations, such as detector response times, limited resolution bandwidth capability when compared to the emission bandwidth, sensitivity, etc., so as to obtain a true peak measurement conforming to the above definitions for the emission in question.

(5) The peak power spectral density is measured as a conducted emission by direct connection of a calibrated test instrument to the equipment under test. If the device cannot be connected directly, alternative techniques acceptable to the Commission may be used. Measurements are made over a bandwidth of 1 MHz or the 26 dB emission bandwidth of the device, whichever is less. A resolution bandwidth less than the measurement bandwidth can be used, provided that the measured power is integrated to show total power over the measurement bandwidth. If the resolution bandwidth is

approximately equal to the measurement bandwidth and
much less than the emission bandwidth of the equipment
under test, the measured results shall be corrected to account
for any difference between the resolution bandwidth of the
test instrument and its actual noise bandwidth.

(6) The ratio of the peak excursion of the modulation envelope
(measured using a peak hold function) to the peak transmit
power (measured as specified above) shall not exceed 13 dB
across any 1 MHz bandwidth or the emission bandwidth,
whichever is less.

(b) Undesirable emission limits: Except as shown in paragraph
(b)(6) of this Section, the peak emissions outside of the frequency
bands of operation shall be attenuated in accordance with the
following limits:

(1) For transmitters operating in the 5.15–5.25 GHz band: all
emissions outside of the 5.15–5.35 GHz band shall not exceed
an EIRP of −27 dBm/MHz.

(2) For transmitters operating in the 5.25–5.35 GHz band: all
emissions outside of the 5.15–5.35 GHz band shall not exceed
an EIRP of −27 dBm/MHz. Devices operating in the 5.25–
5.35 GHz band that generate emissions in the 5.15–5.25 GHz
band must meet all applicable technical requirements for
operation in the 5.15–5.25 GHz band (including indoor use)
or alternatively meet an out-of-band emission EIRP limit of
−27 dBm/MHz in the 5.15–5.25 GHz band.

(3) For transmitters operating in the 5.725–5.825 GHz band: all
emissions within the frequency range from the band edge to
10 MHz above or below the band edge shall not exceed an
EIRP of −17 dBm/MHz; for frequencies 10 MHz or greater
above or below the band edge, emissions shall not exceed an
EIRP of −27 dBm/MHz.

(4) The above emission measurements shall be performed using
a minimum resolution bandwidth of 1 MHz. A lower
resolution bandwidth may be employed near the band edge,
when necessary, provided the measured energy is integrated
to show the total power over 1 MHz.

(5) Unwanted emissions below 1 GHz must comply with the general field strength limits set forth in Section 15.209. Further, any U-NII devices using an AC power line are required to comply also with the conducted limits set forth in Section 15.207.

(6) The provisions of § 15.205 of this Part apply to intentional radiators operating under this Section.

(7) When measuring the emission limits, the nominal carrier frequency shall be adjusted as close to the upper and lower frequency block edges as the design of the equipment permits.

(c) The device shall automatically discontinue transmission in case of either absence of information to transmit or operational failure. These provisions are not intended to preclude the transmission of control or signaling information or the use of repetitive codes used by certain digital technologies to complete frame or burst intervals. Applicants shall include in their application for equipment authorization a description of how this requirement is met.

(d) Any U-NII device that operates in the 5.15–5.25 GHz band shall use a transmitting antenna that is an integral part of the device.

(e) Within the 5.15–5.25 GHz band, U-NII devices will be restricted to indoor operations to reduce any potential for harmful interference to co-channel MSS operations.

(f) U-NII devices are subject to the radio frequency radiation exposure requirements specified in § 1.1307(b), § 2.1091, and § 2.1093 of this Chapter, as appropriate. All equipment shall be considered to operate in a "general population/uncontrolled" environment. Applications for equipment authorization of devices operating under this Section must contain a statement confirming compliance with these requirements for both fundamental emissions and unwanted emissions. Technical information showing the basis for this statement must be submitted to the Commission upon request.

(g) Manufacturers of U-NII devices are responsible for ensuring frequency stability such that an emission is maintained within the band of operation under all conditions of normal operation as specified in the user manual.

GLOSSARY

Access point (AP)   An IEEE 802.11 station that provides distribution services to other stations. APs provide authentication, deauthentication, association, disassociation, reassociation, distribution, *Medium Access Control service data unit* (MSDU) delivery, integration, and privacy services.

Acknowledgement (ACK)   An acknowledgement that data has been received successfully. An ACK message is used with *Return to Send* (RTS) and *Clear to Send* (CTS) messages as part of the MAC *distributed coordination function* (DCF) virtual carrier sense function.

Ad hoc network   A self-contained group of stations communicating directly with each other in a peer-to-peer network. Generally formed spontaneously and existing for a limited time, an IEEE 802.11 ad hoc network is synonymous with an *independent basic service set* (IBSS). An ad hoc network does not include an AP or *distribution system* (DS).

Advanced Encryption Standard (AES)   AES is a cryptographic standard proposed by the *National Institute of Standards and Technology* (NIST). It uses key sizes of 128, 192, and 256 bits.

Association service   The station service that enables a station to map itself to an AP.

Authentication service   The station service that enables a station to establish an identity recognized by other stations. Authentication services are provided by each station. Open system authentication enables any station to become authenticated. Shared key authentication uses a shared secret key. In IEEE 802.11, this key is implemented through the use of *Wired Equivalency Protocol* (WEP) encryption.

Basic service set (BSS)   A set of stations that communicate with each other. BSS variations are the *independent basic service set* (IBSS) or ad hoc network, the *extended service set* (ESS), and the infrastructure BSS. Because the infrastructure service set is the most common type of IEEE *wireless local area network* (WLAN), the acronym BSS is commonly used to specifically refer to an infrastructure basic service set.

Basic service set ID (BSSID) One of the address fields in a MAC frame. The BSSID is a 48-bit field uniquely identifying a BSS. In an infrastructure BSS, it is the IEEE 802.11 MAC address of the AP. In an IBSS, it is a random number calculated by the participating stations.

Binary phase-shift keying (BPSK) See *differential binary phase-shift keying* (DBPSK).

Bluetooth Bluetooth is a wireless *personal area network* (PAN) technology developed by L.M. Ericsson in 1994. It is maintained by an industry group called the Bluetooth *Special Interest Group* (SIG). Bluetooth is a 2.4 GHz technology optimized for a large number of devices communicating in a peer-to-peer relationship.

Broadcast message A multicast message that is addressed to all stations.

Carrier sense multiple access with collision avoidance (CSMA/CA) CSMA/CA is a basic medium access protocol in which network nodes listen before accessing the medium. This mechanism minimizes the amount of data collisions that occur when the medium may be accessed by two nodes simultaneously. Because IEEE 802.11 radio transmission is half duplex, IEEE 802.11 uses CSMA/CA to manage medium access.

Carrier sense multiple access with collision detection (CSMA/CD) CSMA/CD is the access method used in 803.3 Ethernet LANs. When a collision occurs, nodes back off and wait a random amount of time before retrying.

Clear channel assessment (CCA) A function of the *physical* (PHY) layer, which determines the state of the wireless medium (clear, occupied, and so on). A key capability that is utilized by the MAC layer carrier sense mechanism for avoiding collisions during data transmission.

Clear to Send (CTS) A message used in the MAC DCF virtual carrier sense function that indicates the intent to send data over the wireless medium.

Complementary code keying (CCK) CCK is a modulation technique used to provide 5.5 and 11 Mbps transmission rates in IEEE 802.11b. CCK uses nearly orthogonal signal codes and a complex set of Walsh/Hadamard functions known as complementary codes.

Coordination function (CF) The function that manages communication between stations. The CF determines when stations are clear to transmit on the wireless medium. There are two types of CFs in the IEEE 802.11 MAC: the *distributed coordination function* (DCF) and the *point coordination function* (PCF).

Data Encryption Standard (DES) DES is a cryptographic standard adopted in 1976 by the NIST. DES encrypts and decrypts data using a 64-bit key with an actual length of 56 bits due to the use of 1 parity bit included in each group of 8 bits.

Deauthentication service The station service that enables a station to terminate an existing authentication.

Differential binary phase-shift keying (DBPSK) DBPSK is a basic *direct sequence spread spectrum* (DSSS) modulation technique in which the binary (0,1) data signal is determined by two signal phase shifts (0° and 180°). It is used for 1 Mbps transmission in IEEE 802.11.

Differential quadrature phase-shift keying (DQPSK) DQPSK is a *direct sequence spread spectrum* (DSSS) modulation technique in which the binary (0,1) data signal is determined using four signal phase shifts (0°, 90°, 180°, and 270°). It is used for 2 Mbps transmission in IEEE 802.11.

Digital Enhanced Cordless Telephone (DECT) The DECT standard is designed to provide wireless communications for telecommunications equipment such as cordless phones and *private branch exchanges* (PBXs). It operates in the 1.9 GHz radio band and employs *Time Division Multiple Access* (TDMA) technology. It has been in existence since 1992 and is maintained by the *European Telecommunications Standards Institute* (ETSI).

Direct sequence spread spectrum (DSSS) DSSS is a spectrum management technique that spreads the signal over a larger frequency band by multiplexing each bit with a fast repetitive pattern called a *code*.

Disassociation service   The station service that enables a station to terminate an association relationship.

Distributed coordination function (DCF)   A type of coordination function in which all stations in a BSS use the same mechanism for determining the availability of the medium for transmission. This is a protocol in which stations contend for medium access. This is compared with the PCF, which is a contention-free protocol. The DCF of the IEEE MAC is implemented using CSMA/CA.

Distributed interframe space (DIFS)   The DIFS is used by the DCF collision avoidance mechanism for transmission of data and management frames.

Distribution service   The station service that enables a station to deliver messages between APs in an ESS.

Distribution system (DS)   The communication system for a set of interconnected BSSs forming an ESS. This is generally implemented on an 802.3 LAN.

Dynamic Host Control Protocol (DHCP)   DHCP is an *Internet Protocol* (IP) network protocol that enables configuration information to be passed from a server to a host. It also enables the assignment of network addresses to the host.

Extended interframe space (EIFS)   The EIFS is used by the DCF for processing frames received with an incorrect *frame check sequence* (FCS).

Extended service set (ESS)   A network of infrastructure BSSs with their APs connected by a DS (usually an 802.3 LAN). Stations within an ESS may move from BSS to BSS through the use of the reassociation service. This movement is transparent to entities communicating to ESS stations from external networks.

Frequency-hopping spread spectrum (FHSS)   FHSS is a spectrum management technique that spreads the signal over a larger frequency band by allocating channels across the frequency band and transmitting on different channels in a pattern.

Frequency-shift keying (FSK) See *Gaussian frequency-shift keying* (GFSK).

Gaussian frequency-shift keying (GFSK) A modulation scheme used in the IEEE 802.11 FHSS PHY. Two variations are specified in IEEE 802.11. *Two-level GFSK* (2GFSK) provides 1 Mbps FHSS operation. *Four-level GFSK* (4GFSK) provides 2 Mbps FHSS operation. A Gaussian filter is applied to the baseband signal. In 2GFSK, a binary 1 is represented by a positive deviation from the base carrier and a 0 is represented by a negative deviation. 4GFSK represents four values (00,01,10,11) by using four carrier deviations.

High-performance radio local area network (HIPERLAN) HIPERLAN is a European standard for broadband WLANs. It is maintained by the ETSI.

Independent basic service set (IBSS) A self-contained group of stations communicating directly with each other in a peer-to-peer network. Generally formed spontaneously and existing for a limited time, an IEEE 802.11 IBSS is synonymous with an ad hoc network. An IBSS does not include an AP or DS.

Infrastructure basic service set (BSS) A network of stations communicating with each other through the coordination function of an AP. As the most common form of a basic service set, it is sometimes referred to as a BSS. An infrastructure BSS may communicate with other networks through the portal function of an AP.

Integration service The station service that enables a station to transfer messages outside of the WLAN. This is sometimes called *portal* or *gateway functionality*.

Interframe space (IFS) This is the time interval between IEEE 802.11 frames. There are four types of IFSs: the short IFS, the DCF IFS, the PCF IFS, and the extended IFS.

Internet service provider (ISP) A company that provides a connection to the Internet. Connectivity ranges from dial up at several hundred bits per second to very high-speed optical. See *wireless ISP* (WISP).

Local area network (LAN)   A network covering an area associated with an office or campus environment. The most common type of LAN is based on the IP suite.

MAC management protocol data unit (MMPDU)   A message exchanged between two MAC entities (for example, the MAC layers present in two stations) for the transfer of MAC management information.

MAC protocol data unit (MPDU)   A message exchanged between two MAC entities (for example, the MAC layers present in two stations) for the transfer of data.

MAC service data unit (MSDU)   A message exchanged between LLC entities that utilize the MAC layer for delivery.

Medium Access Control (MAC)   The MAC layer is the lower sublayer that IEEE 802.11 specifications provide for managing access to the PHY. Combined with the *Logical Link Control* (LLC) layer, it is equivalent to the *Open Systems Interconnection* (OSI) data link layer.

Metropolitan area network (MAN)   A MAN is a network covering an area larger than a LAN, but smaller than a *wide area network* (WAN). A MAN is generally applied to high-density networks used in metropolitan areas. The IEEE 802.16 Working Group maintains specifications for MANs.

MSDU delivery service   The station service that enables a station to transmit messages to another station.

Multicast message   A message addressed to multiple stations using the group bit in the MAC address field.

Network allocation vector (NAV)   A time value maintained by a station as part of its DCF function that indicates when the wireless medium will be free for a transmission. The NAV is part of the CSMA/CA mechanism.

Open Systems Interconnection (OSI) network model   The *International Standards Organization* (ISO) has specified a layered network model that serves as the model for many network designs. It consists of seven layers: physical, data link, network, transport, session, presentation, and application. The IEEE 802.11 standards specify the physical layer (IEEE 802.11 PHY) and the lower half of the data link layer (IEEE 802.11 MAC).

Packet binary convolutional coding (PBCC)   PBCC is an optional modulation technique used in IEEE 802.11b and 802.11g.

Personal area network (PAN)   A PAN generally covers an area less than 10 meters. PAN technologies include wireless standards such as Bluetooth. The IEEE 802.15 Working Group maintains standards for PANs.

Phase-shift keying (PSK)   A modulation technique where binary data values (0,1) are determined by evaluating the phase shift of a signal. See *binary phase-shift keying* (BPSK) and *quadrature phase-shift keying* (QPSK).

Physical (PHY) layer   The lowest layer in the OSI protocol model. The PHY layer in IEEE 802.11 provides the interface to the physical transmission media for the MAC layer. It consists of the *PHY Layer Convergence Procedure* (PLCP) sublayer and the *PHY medium dependent* (PMD) sublayer.

PHY Layer Convergence Protocol (PLCP)   This is the sublayer of the IEEE 802.11 PHY that translates frames received from the MAC into information understood by the PMD.

PHY medium dependent (PMD) layer   This is the sublayer of the IEEE 802.11 PHY that interfaces directly to the physical medium. It provides medium-specific functions such as sensing the medium for carrier.

Point coordination function (PCF)   The PCF is a coordination function implemented in an AP providing contention-free (dedicated) access to the wireless medium. It is used for services sensitive to delay and latency. The PCF provides a station with dedicated access to the medium.

Point coordinator (PC)   A logical function within an AP that determines when a station may transmit.

Point interframe space (PIFS)   The PIFS is used by the PCF to obtain exclusive access to the wireless medium after the medium is determined to be idle.

Portal A point where service messages to/from other networks leave/enter an IEEE 802.11 network. This is also called a *gateway*. Portal functions may be implemented in an AP.

Privacy service The station service that enables a station to encrypt the contents of messages. In IEEE 802.11, this is accomplished through the use of the WEP encryption algorithm. Privacy services are not used for all messages. The initial negotiation of authentication and privacy services between stations is sent in the clear.

Quadrature phase-shift keying (QPSK) See *differential quadrature phase-shift keying* (DQPSK).

Reassociation service The station service that enables a station to move its association from one BSS to another. It is also used to change its attributes within the same BSS.

Shared Wireless Access Protocol (SWAP) SWAP is a specification developed by the HomeRF organization for wireless communication in the 2.4 GHz band. SWAP supports voice as well as data and is targeted for use by consumers in the home.

Short interframe space (SIFS) SIFS is used when the station requires the wireless medium for the duration of a frame exchange sequence such as *Clear to Send* (CTS) to data to *acknowledgment* (ACK).

Station A device that utilizes the IEEE 802.11 MAC and PHY for network connectivity. Stations may be fixed (don't move), portable (move, but only connect to the network when stationary), or mobile (connect to the network while moving). A station providing distribution services is an AP.

Unicast message A message with a single recipient. This is also called a *directed message*.

Virtual carrier sense A mechanism for determining if the wireless medium is being used for a transmission. Part of the MAC DCF CSMA/CA mechanism.

Virtual private network (VPN) A network access method that provides a high level of data security over a shared or easily accessible transmission media. A VPN's security features may include encryption, authentication, and tunneling.

Wide area network (WAN)   A network that spans a larger area relative to a LAN (office/campus) or MAN.

Wired Equivalent Protocol (WEP)   The optional cryptographic algorithm used to provide privacy in wireless data transmissions. The WEP provides a minimum amount of privacy and is vulnerable to attack. This is usually supplemented with a VPN or other security mechanisms for networks requiring a higher level of security.

Wireless Application Protocol (WAP)   A mobile device protocol that provides Internet access for a variety of Internet-based applications.

Wireless ISP (WISP)   An ISP using wireless media (radio and laser) for connectivity to the Internet.

INDEX